Recent Studies in Atomic and Molecular Processes

PHYSICS OF ATOMS AND MOLECULES

ATOM – MOLECULE COLLISION THEORY: A Guide for the Experimentalist
Edited by Richard B. Bernstein

ATOMIC INNER-SHELL PHYSICS
Edited by Bernd Crasemann

ATOMS IN ASTROPHYSICS
Edited by P. G. Burke, W. B. Eissner, D. G. Hummer, and I. C. Percival

AUTOIONIZATION: Recent Developments and Applications
Edited by Aaron Temkin

COHERENCE AND CORRELATION IN ATOMIC COLLISIONS
Edited by H. Kleinpoppen and J. F. Williams

COLLISIONS OF ELECTRONS WITH ATOMS AND MOLECULES
G. F. Drukarev

DENSITY MATRIX THEORY AND APPLICATIONS
Karl Blum

ELECTRON AND PHOTON INTERACTIONS WITH ATOMS
Edited by H. Kleinpoppen and M. R. C. McDowell

ELECTRON-ATOM AND ELECTRON-MOLECULE COLLISIONS
Edited by Juergen Hinze

ELECTRON-MOLECULE COLLISIONS
Edited by Isao Shimamura and Kazuo Takayanagi

INNER SHELL AND X-RAY PHYSICS OF ATOMS AND SOLIDS
Edited by Derek J. Fabian, Hans Kleinppopen, and Lewis M. Watson

INTRODUCTION TO THE THEORY OF LASER-ATOM INTERACTIONS
Marvin H. Mittleman

ISOTOPE SHIFTS IN ATOMIC SPECTRA
W. H. King

PROGRESS IN ATOMIC SPECTROSCOPY, Parts A, B, C, and D
Edited by W. Hanle, H. Kleinpoppen, and H. J. Beyer

RECENT STUDIES IN ATOMIC AND MOLECULAR PROCESSES
Edited by Arthur E. Kingston

THEORY OF MULTIPHOTON PROCESSES
Farhad H. M. Faisal

VARIATIONAL METHODS IN ELECTRON-ATOM SCATTERING THEORY
R. K. Nesbet

A Continuation Order Plan is available for this series. A continuation order will bring delivery of each new volume immediately upon publication. Volumes are billed only upon actual shipment. For further information please contact the publisher.

Recent Studies in Atomic and Molecular Processes

Edited by

Arthur E. Kingston
The Queen's University of Belfast
Belfast, Northern Ireland

PLENUM PRESS • NEW YORK AND LONDON

Library of Congress Cataloging in Publication Data

Recent studies in atomic and molecular processes.

(Physics of atoms and molecules)
"Proceedings of a conference held in honor of the 70th birthday of Professor Sir David Bates, held November 17-18, 1986, at the Queen's University of Belfast, Belfast, Northern Ireland"—T.p. verso.
Includes index.
1. Atoms—Congresses. 2. Molecules—Congresses. 3. Bates, David R. (David Robert), 1916- . I. Kingston, Arthur E. II. Series.
QC172.R43 1987 539'.7 87-7263

ISBN-13: 978-1-4684-5400-0 e-ISBN-13: 978-1-4684-5398-0
DOI: 10.1007/978-1-4684-5398-0

Proceedings of a conference held in honor of the 70th birthday of Professor Sir David Bates, held November 17-18, 1986, at the Queen's University of Belfast, Belfast, Northern Ireland

Softcover reprint of the hardcover 1st edition 1987
A Division of Plenum Publishing Corporation
233 Spring Street, New York, N.Y. 10013

PREFACE

Sir David Bates celebrated his seventieth birthday on the 18th November 1986. To mark this event a conference was held in the David Bates building at The Queen's University of Belfast on the 17th and 18th November 1986.

At this conference ex-students and colleagues of Sir David, who are acknowledged world experts in their field of research, gave in depth reviews of a particular area of atomic and molecular physics. This book on the conference presents a unique account of recent studies in atomic and molecular processes in a wide range of research fields.

This volume is dedicated to Sir David by his friends as a token of their affection and respect. It is hoped that it will provide a useful summary of current research in atomic and molecular physics and that it will also show the great contribution which Sir David made to atomic and molecular physics.

This conference was supported by the USAF European Office of Aerospace Research and Development who we wish to thank for their generous support.

Arthur E. Kingston

CONTENTS

INTERSTELLAR CLOUD CHEMISTRY REVISITED

David R. Bates

Department of Applied Mathematics & Theoretical Physics
Queen's University of Belfast
Belfast BT7 1NN, U.K.

INTRODUCTION

My first encounter with interstellar cloud chemistry (Bates and Spitzer 1951) took place 35 years ago. The field then had a semblance of simplicity. Absorption lines in the spectra of a number of stars had revealed the existence of molecules in interstellar space. The only identified neutrals were CH and CN (Swings and Rosenfeld 1937, McKellar 1940, Adams 1941) and the only identified ion was CH^+ (Douglas and Herzberg 1941).

Spitzer and I tried to explain the equilibrium CH and CH^+ abundances. We judged it should be sufficient to take into account their formation by radiative association

$$C + H \rightarrow CH + h\nu \quad , \quad C^+ + H \rightarrow CH^+ + h\nu \tag{1}$$

and their destruction by photo-dissociation, photo-ionization and dissociative recombination

$$CH^+ + e \rightarrow C + H \tag{2}$$

which a recent theoretical development (Bates 1950) had correctly led us to regard as a major sink. Failing in our endeavour we considered introducing ion-molecule reactions like

$$H_2 + C^+ \rightarrow H + CH^+ \tag{3}$$

but because of the severe limitation on the range of reactants it then seemed reasonable to invoke found none which are exothermic. We also considered the possibility of CH and CH^+ being formed on grains. Dalgarno (1976) has given a review of the CH, CH^+ problem. Later work on it has been outlined by Duley and Williams (1984). There is a recent paper by Pineau des Forêts et al (1986).

The wonderful development of the subject has made current interstellar cloud chemistry far more complicated than it was in its nonage. One measure of this is the list of 90 identified interstellar molecules compiled by Mann and Williams (1980). Another is an investigation by Huntress and Prasad (1980) using a computer code that involves 137 species and allows for over 1400 reactions. Because of the tyranny of age it would be vain for me to attempt to re-enter properly so complex a field of research with moreover a vast literature. Happily the periphery of the field has not been fully explored. I will outline what I have been doing recently in the shallows around the edge.

HITTING COLLISIONS BETWEEN IONS AND POLAR OR QUADRUPOLAR MOLECULES

The rate coefficient for hitting (or intimate) collisions between an ion and a molecule is important because, if an exothermic channel exists, it is usually a good approximation to the rate coefficient for the exothermic process. It (or the corresponding cross section) is also needed in calculations on radiative association.

In the case of a molecule of isotropic polarizability α that has no permanent dipole or quadrupole moment the hitting rate coefficient k_L is given (cf Eyring, Herschfelder and Taylor 1936, Gioumousis and Stevenson 1958) by

$$k_L = 2\pi(\alpha/\mu)^{\frac{1}{2}} \tag{4}$$

where μ is the reduced mass of the colliding pair.

The quantal perturbed rotating state (PRS) approximation of Takayanagi (1978) was used by himself (Takayanagi 1979, 1982a,b) and by Sakimoto (1980) to calculate rate coefficients $k(J)$ for hitting collisions between ions and some polar and quadrupolar molecules. It was noted that when the rotational quantum number J of the molecule and the kinetic temperature T are small the ratio $k(J)/k_L$ may be very large and stress was laid (Sakimoto 1980, Takayanagi 1982c) on the relevance of this in interstellar cloud chemistry. The attention of the general astrophysical community was not caught.

Prior to the Japanese work theorists concerned with ion-molecule reactions in the laboratory had applied classical methods to the problem of collisions in which the interaction is orientation dependent (cf Su and Bowers 1975, 1979). They were not interested in the hitting rate coefficient for the different J levels. A method called ADO (or AQO), for average dipole orientation (or average quadrupole orientation), appealed to me physically. In it the interaction appropriate to an average orientation of the rapidly rotating molecule was adopted. After recognizing that the averaging was flawed I introduced an overlooked facet of the collision, adiabatic invariance (Bates 1982). The hope was that this would enable a sufficiently accurate calculation of $k(J)/k_L$ to be carried out much more easily than by the PRS approximation of Takayanagi (1979). With the help of a key contribution by Sakimoto (1984) this hope has been realized (Bates and Mendaš 1985, Morgan and Bates 1986, Bates and Morgan 1986).

The rate of change of the orbital radial distance r is slow in the sense that if τ is the time period of the rotation (or libration) of the molecule then

$$\tau \frac{dr}{dt} << r \tag{5}$$

throughout the encounter. Letting ε be the sum of the kinetic energy of rotation (or libration) and the potential energy in the orientation dependent part of the field we have that since r varies slowly $d\varepsilon/dt$ must be proportional to $F(r)\, dr/dt$, $F(r)$ being a function of r and therefore that there is some combination of ε and r that remains constant. Using Hamilton's equations of motion it may be shown (Landau and Lifshitz 1960) that this adiabatic invarient combination is

$$I = \oint p_\theta \, d\theta/\pi \tag{6}$$

where θ is the polar angle, p_θ is the conjugate momentum and the integration is carried out over a complex rotation (or libration). Use of eq (6) enables ε to be determined provided (ϕ,ψ) the remaining two Eulerian angles are cyclic.

Consider as an example (Bates and Morgan 1987) the case of a symmetrical top of rotational constants A and B having a dipole of moment D along its axis. We have that

$$\varepsilon(r) = De\cos\theta/r^2 + (\hbar^2/4)\,[\{\dot\theta^2 + \dot\phi^2\sin^2\theta\}/B + \{\dot\psi + \dot\phi\cos\theta\}^2/A] \quad (7)$$

The conjugate momenta are given by

$$p_\theta = \dot\theta\,\hbar^2/2B \quad (8)$$

$$p_\phi = \dot\phi\,\hbar^2(B\cos^2\theta + A\sin^2\theta)/2AB + \dot\psi\hbar^2\cos\theta/2A = m\hbar \text{ (say)} \quad (9)$$

$$p_\psi = \hbar^2(\dot\psi + \dot\phi\cos\theta)/A = K\hbar \text{ (say)} \quad (10)$$

m and K being constants because of ϕ and ψ being cyclic coordinates. Using eq. (8) - (10) to eliminate $\dot\theta$, $\dot\phi$ and $\dot\psi$ from eq. (7) we get

$$\varepsilon(r) = De\cos\theta/r^2 + \{Bp_\theta^2/\hbar^2 + B(m - K\cos\theta)^2/\sin^2\theta + AK^2\} \quad (11)$$

Change the variable r to the dimensionless x where

$$De\,x^2 = Br^2 \quad (12)$$

put

$$\cos\theta = \rho \quad (13)$$

and introduce

$$v(x) = \{\varepsilon(r) - \varepsilon(\infty)\}/B \quad (14)$$

with

$$\varepsilon(\infty) = B(J + \tfrac{1}{2})^2 + (A - B)K^2 \quad (15)$$

Eq. (11) then gives

$$p_\theta = \hbar\,\{v(x) + (J + \tfrac{1}{2})^2 - K^2 - \rho\bar{x}^2 - (m - K\rho)^2/(1 - \rho^2)\}^{\frac{1}{2}} \quad (16)$$

and eq. (6) may be written

$$I = \frac{\hbar}{x\pi}\int \frac{f(\rho)^{\frac{1}{2}}\,d\rho}{1 - \rho^2} \quad (17)$$

with

$$f(\rho) \equiv \rho^3 - \rho^2 x^2\,[v(x) + (J + \tfrac{1}{2})^2] - \rho[1 - 2mKx^2] + x^2[v(x) + (J + \tfrac{1}{2})^2 - m^2 - K^2]\,. \quad (18)$$

Integral (17) may be evaluated without difficulty. In the $x \to \infty$ limit the evaluation gives

$$I = (J + \tfrac{1}{2} - |\ell|)/\hbar \quad (19)$$

where $|\ell|$ is the larger of $|K|$ and $|m|$ while at finite x it gives a cumbersome expression for I in terms of the complete elliptic integrals of the first, second and third kinds. This expression and eq. (19) enable v(x) to be obtained by iteration. By differentiating eq. (17) a relatively simple expression for dv(x)/dx may be derived. The completion of the problem is now straightforward. Letting η be the incident energy of relative motion in units B and b be the impact parameter in units $(De/B)^{\frac{1}{2}}$ we see on adding the ion-induced dipole interaction and centrifugal energy to v(x) that the effective potential for the encounter is

$$V(eff) = -Px^{-4} + \eta b^2 x^{-2} + v(x) \tag{20}$$

with

$$P \equiv \alpha B/2D^2 \tag{21}$$

At the maximum of V(eff) it is evident from eq. (20) that

$$\eta b^2 = 2Px^{-2} + \frac{1}{2} x^3 \frac{dv}{dx} . \tag{22}$$

The energy of relative motion just great enough to traverse the maximum is

$$\eta = V(eff) + \frac{1}{2}x \frac{dV(eff)}{dx} \tag{23}$$

from which terms involving b cancel to give

$$\eta = Px^{-4} + \frac{1}{2} x \frac{dv}{dx} + v(x) . \tag{24}$$

If $x(\eta)$ is obtained by inverse interpolation from eq. (24) substitution in eq. (22) yields $b(\eta)$ from which the hitting rate coefficient k(J,K) may readily be calculated. Note that the rate coefficient does not depend on A, this rotational constant being eliminated when v(x) of eq. (14) is chosen as the dependent variable.

The analysis for the symmetrical top covers a linear molecule. All that need be done is to set K = 0 throughout (Morgan and Bates 1987).

The adiabatic invarience method may be used for collisions involving any orientation dependent interaction that may be expressed as a simple function of $\cos\theta$: for example the case of the quadrupolar molecule,

$$U = -\alpha e^2/2r^4 + eq\, P_2(\cos\theta)/r^3 \tag{25}$$

or the case of a molecule having an anisotropic polarizability

$$U = -\bar{\alpha}e^2 \{1 + 2KP_2(\cos\theta)\}/2r^4 \tag{26}$$

in which

$$\bar{\alpha} = \frac{1}{3} \{\alpha(||) + 2\alpha(\perp)\} , \quad K = \{\alpha(11) - \alpha(\perp)\}/3\bar{\alpha} \quad . \tag{27}$$

Calculations have been done using interaction (25) and taking the molecule to be linear, (Bates and Mendas 1985).

Fig. 1 shows graphs of $k(J)/k_L$ against T (with J = 0 or 1) for a representative linear polar molecule HCl as calculated by the adiabatic invarience method (Morgan and Bates 1987). The ratios are large especially when T is low. Values obtained by Sakimoto (1980) using the accurate PRS method are also shown as are values obtained by Adams, Smith and Clary (1985) using a quantal method (AC) developed by Clary (1985) that introduces localized rotational basis functions in order to speed the convergence. All the results are in excellent agreement above 20K

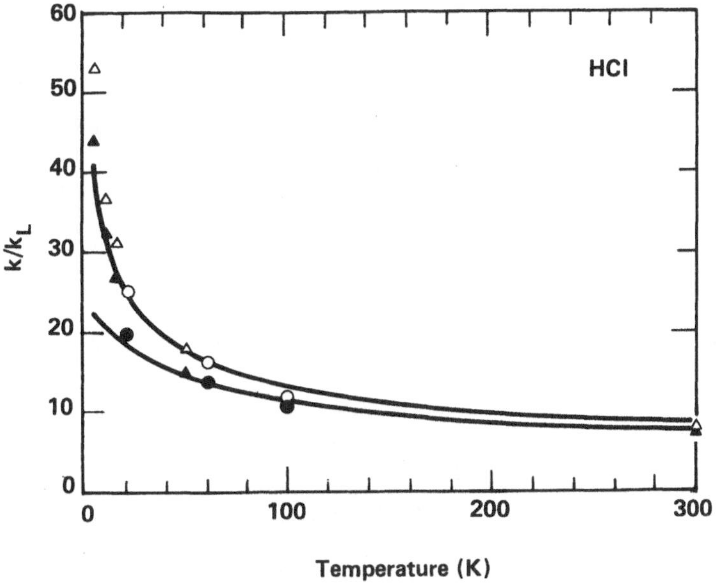

Fig. 1. Comparison of adiabatic invarience results, solid lines, with results of other calculations. Circles are PRS results of Sakimoto (1980), triangles are AC results of Adams et al (1985); the upper line and open symbols refer to J = 0 and the lower line and filled symbols refer to J = 1 (after Morgan and Bates 1987).

and Clary, Smith and Adams (1985) have checked with laboratory data above 205K. The agreement between the adiabatic invarience and PRS results continues at lower temperatures but the AC results become rather high (see Morgan and Bates 1987). Having carried through an extensive set of computations Morgan and Bates (1987) parameterized their results so that both cross sections and rate coefficients for any polarizability, dipole moment, reduced mass and temperature can easily be got.

The k(K,1) rate coefficients for a polar symmetrical top are rather greater than the k(1,0) rate coefficients (which are of course identical with those for the corresponding linear molecule). They too have been parameterized. (Bates and Morgan 1987).

Using their CRESU technique Marquette et al (1985) have measured the rate coefficients k(rc) for reactive $C^+ - NH_3$ and $N^+ - NH_3$ collisions at 27K and 68K. Combination of the CRESU values of k(rc) and the theoretical values of k leads to collision efficiencies k(rc)/k of 0.6 to 0.7 (Bates and Morgan 1987). These quite high collision efficiencies are in harmony with recent work by Herbst (1986a) on the question of whether the large predicted rate coefficients might not be greatly reduced by short-range centrifugal barriers to reaction. Herbst concluded that such barriers are generally important only for reactions that are slow at 300K.

Quadrupole moments have less effect on the hitting rate coefficients than have polar moments. Figs. 2 and 3 show some of the linear molecule results of Bates and Mendas (1985). The parameter κ which occurs in the caption is defined by

$$\kappa = 9.67 \times 10^{-2} \; \tilde{\alpha} \; \tilde{B}^{\frac{1}{2}} \; |\tilde{q}|^{-4/3} \qquad (28)$$

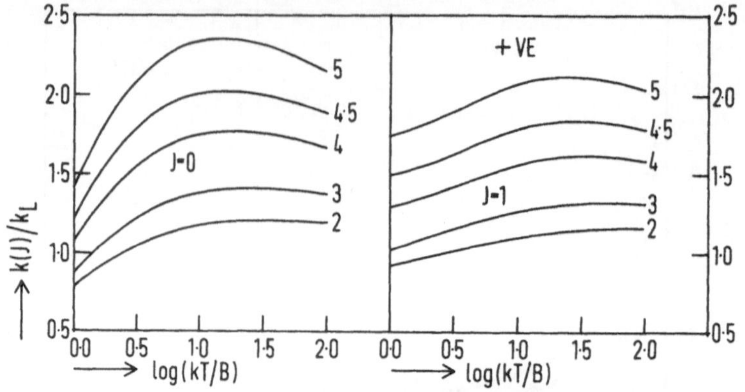

Fig. 2. Adiabatic invarience results on $k(J)/k_L$ for eq positive as function of $\ell g(kT/B)$ with J for each set of five curves as indicated and with $\ell g(\kappa^{-3})$ also as indicated where κ is given by eq. (28) (after Bates and Mendaš 1985).

where the tilde symbol placed over the polarizability α, the quadrupole moment q and the rotational constant B signifies that these entities are expressed in the usual practical units ($10^{-24}cm^3$, $10^{-26}cm^2$ e.s.u. and cm^{-1} respectively). It will be noticed that the rate coefficients for eq positive are less than those for eq negative. This is in accord with what one would expect from some PRS results depicted by Takayanagi (1982b). However Bhowmik and Su (1986) reached the opposite conclusion from some trajectory calculations intended to give the rate coefficients for selected molecules at 300K averaged over the contributing rotational levels. Bates and Mendaš (1985) presented thermal average results in the form of graphs having a parameter related to κ of eq (28) as the independent variable.

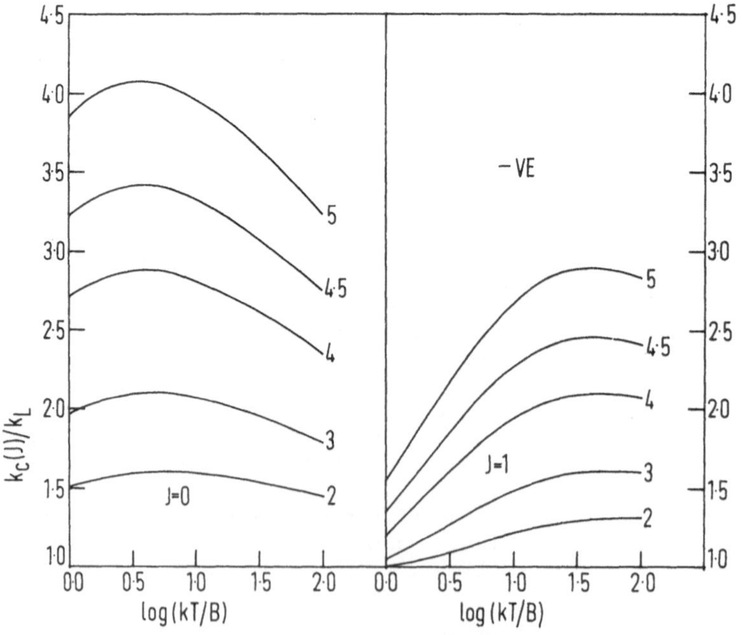

Fig. 3. The same as fig. 2 except that eq is negative

The trajectory analysis results for eq positive lie satisfactorily along the appropriate graph but those for eq negative are scattered.

After astrophysicists came to realize that with polar molecules $k(J)/k_L$ is large for low J and low T enquires on consequences were naturally begun.

Millar et al (1985) took up the problem of explaining the curiously large $[HCS^+]/[CS]$ ratio in interstellar clouds. As they noted one factor is the smallness of $\alpha_D(HCS^+)$ the dissociative recombination coefficient measurements which they made giving about 5×10^{-8} cm^3 s^{-1} at 300K. They also noted that CS has a large proton affinity so that HCS^+ can be destroyed in collisions with only a few species of neutrals whereas it may be formed in collisions between CS and many species of ions for example

$$H_3^+ + CS \rightarrow HCS^+ + H_2 \ . \tag{29}$$

Moreover radiative transitions between the rotational levels are fast enough to ensure that the CS is in the J = 0 level that is in the level for which the hitting collision rate coefficient is greatest. Applying their scheme Millar et al (1985) reproduced the $[HCS^+]/[CS]$ ratio satisfactorily.

Herbst and Leung (1986) have run pseudo time dependent models of the gas phase chemistry of dense interstellar clouds with the large predicted rate coefficients for ion-molecule reactions involving polar molecules. They found that the incorporation of the large rate coefficients normally lead to a reduction in both the peak and steady state abundances which can be as great as by a power of ten but is commonly smaller.

DISSOCIATIVE RECOMBINATION OF POLYATOMIC IONS

Erroneous research is happily soon forgotten. But Oblivium is fairly even-handed in its treatment of individuality be it bad or good in science. Most of us sadly come to recognize that even our best pieces of research have only brief life-times in recognizable forms. As I turned to dabbling in interstellar cloud chemistry I would therefore have been surprised if I had found that my single page Letter (Bates 1950) describing dissociative recombination had survived. It had not. However I confidently expected to find that its content would be thoroughly understood by all. In this I was grievously disappointed. To be sure everyone knew that dissociative recombination is very rapid - after all its great rapidity has long been established experimentally. To be sure also everyone knew that dissociative recombination involved the break-up of the ion - after all this is just what the name implies. However many of those engaged in interstellar cloud chemistry seemed to think that there is no theory relevant to the dissociative channels and consequently felt free to propose any dissociative channel they pleased. There are numerous illustrations of this in the literature. One is the proposal that the source of cyanoethynyl is

$$H_2C{=}CH{=}CN^+H + e \rightarrow C{\equiv}C{-}C{\equiv}N + 2H_2 . \tag{30}$$

Herbst (1978) deserves considerable credit for being the first astrophysicist to enquire into what are the products of polyatomic ion dissociative recombination. He based his study on the statistical phase space model of Pechukas and Light (1965) which he used to calculate the relative probabilities of the various exoergic dissociative channels. Later Green and he (1979) in another pioneering investigation argued that the rate at which potential energy is converted into kinetic energy is inversely proportional to product mass so that hydrogen atoms are likely to be expelled more rapdily from the complex than are heavy atoms or molecular fragments;

that if there are several hydrogen atoms any one will be expelled with approximately equal probability; and that multi-hydrogen atom expulsion as in

$$H_2C_3O^+ + e \rightarrow C_3O + H_2 \quad (31)$$

and

$$H_3C_3O^+ + e \rightarrow C_3O + H_2 + H \quad (32)$$

(Herbst, Smith and Adams 1984) can also occur. General acceptance of the conclusions of Green and Herbst (1979) appears to be implicit in the work of many interstellar cloud chemists although Dalgarno (1986) expressed a cautious viewpoint. However neither this nor the earlier approach of Herbst (1978) incorporates the essential feature of dissociative recombination (Bates 1986a). My original simple model of the process (Bates 1950) remains useful and I will therefore recall it briefly.

Because of the misbelief that such a process could only take place through a break-down of the Born-Oppenheimer approximation (cf Massey 1950) it was originally presumed that a process like dissociative recombination is slow. Investigations on recombination in the Earth's ionosphere (Bates and Massey 1947) suggested the contrary. After an uncertain start (cf Massey and Gilbody 1974) the application of the microwave probing technique for observing the rate of decay of the electron density in afterglows by Biondi and Brown (1949) proved that dissociative recombination is indeed rapid. Realization then came (Bates 1950) that the process does not entail break-down of the Born-Oppenheimer approximation. It may be regarded as taking place in two stages. First the system comprising the molecular ion and the free electron undergoes a radiationless transition to yield a neutral molecule on a repulsive potential energy surface. Second the atoms of the neutral molecule quickly move apart preventing the inverse radiationless transition and thereby stabilizing the recombination. The radiationless transition is normally the rate-limiting step. Let A be its rate and let g(i) and g(n) be the statistical weights of the states of the ionized and neutral molecule that are involved. Denoting the initial vibrational wave function by $\chi(r)$ and approximating to the final vibrational wave function by the delta function of Winans and Stueckelberg (1928) it may be shown that the dissociative recombination coefficient at electron temperature T is

$$\alpha = h^3 A g(n) F_1 / 2(2\pi mkT)^{3/2} g(i) \quad (33)$$

with

$$F_1 = \int \chi(r)^2 \frac{dr}{d\varepsilon} \exp(-\varepsilon/kT) d\varepsilon \quad (34)$$

in which ε is the energy supplied by the electron in a vertical transition between the two potential surfaces and $d\varepsilon/dr$ is the gradient of the repulsive potential. It is evident that the dimensionless entity F_1 is in the nature of a Franck-Condon factor. Suppose that the ion is in its ground vibrational level, that the intersection of the potential surfaces occurs near where $\chi(r)$ is passing through its maximum and that the local gradient there $(d\varepsilon/dr)_o$ is steep in the sense that a change greatly exceeding kT occurs in the r-range over which $\chi(r)$ is appreciable. Taking $\chi(r)$ to be as for a simple harmonic oscillator we find from eq (34) that

$$F_1 \simeq kT\{2\mu\omega c/\hbar\}^{\frac{1}{2}} / (d\varepsilon/dr)_o \quad (35)$$

where μ is the reduced mass of the pair of atoms and ω is the vibrational frequency. Note that F_1 is a very small number - typically about the magnitude of kT when expressed in eV.

An important feature of dissociative recombination is that a radiationless transition in which a free electron enters an anti-bonding orbital and another electron is simultaneously excited to such an orbital from a bonding orbital may cause the dissociation of even an ion having a strong triple valence bond. The case of

$$NO^+ + e \rightarrow N + O \tag{36}$$

which has been investigated theoretically by Bardsley (1968) is instructive. In the notation of Herzberg (1950), but with subscripts b or a added to indicate whether the orbital is bonding or anti-bonding, the electron configuration of the ion is

$$KK(\sigma 2s_b)^2 \; (\sigma 2s_a)^2 \; (\sigma 2p_b)^2 \; (\pi 2p_b)^4 \quad . \tag{37}$$

The six outer electrons constitute a triple-valence bond. Dissociation would ensue from a radiationless transition to the state of the molecule ($I^2\Sigma^+$ or $B^{1\;2}\Delta$) having the configuration

$$KK(\sigma 2s_b)^2 \; (\sigma 2s_a)^2 \; (\sigma 2p_b) \; (\pi 2p_b)^4 \; (\pi 2p_a)^2 \tag{38}$$

Of the seven outer orbitals five are bonding and only two are anti-bonding. Nevertheless the potential is repulsive in the relevant region (becoming rather weakly attractive at larger separations).

Dissociative recombination of a polyatomic ion is closely similar to dissociative recombination of a diatomic ion. Electronic energy is again converted into energy of relative motion of heavy particles through the interactions that come into play after a radiationless transition at a crossing of potential energy surfaces of the type that has been discussed. Any of the valence bonds belonging to the ionized atom in the polyatomic ion may become a repulsion in consequence of the transition as may an ion-induced dipole attraction.

The contribution to the dissociative recombination coefficient from each of the identical bonds of an ion like CH_3^+ is of course the same. Letting n be the number of identical bonds it is instructive to express experimental data in the form α/n that is the dissociative recombination coefficient per bond. This is done in table 1 in which the value for NH_4^+ is taken from Alge, Adams and Smith (1983) and the remaining values are taken from McGowan et al (1979). The values of α/n at 120K lie within a factor of 2.1 of 2.8×10^{-7} cm^3 s^{-1} for the ions with one or more hydrogen atoms and within a factor of 1.8 of 5.3×10^{-7} cm^3 s^{-1} for the ions with two heavy atoms. The measured α(in units of 10^{-7} cm^3 s^{-1}) for $HC \equiv C^+H$, $H_2C = C^+H$ and $HN^+ \equiv N$ are 9.4, 15 and 12 which values are too high to fit readily with the entries in the first row of table in conjunction with the view that single hydrogen atom ejection is the main channel in the dissociative recombination of polyatomic ions. However the corresponding values obtained by adding the appropriate entries in the first and second rows are 13, 13 and 7 respectively. This represents satisfactory agreement in the present context.

The ease with which multiple bonds can in general be dissolved in dissociative recombination makes the synthesis of complex molecules less efficient than has commonly been assumed in the past.

As implied in the preceding statement there are exceptions. The crossing of the potentials may be unfavourable. The recombination energy may be insufficient to break the multiple bond as in

Table 1. Measured Dissociative Recombination per bond α/n at 120K

ion	CH^+	CH_2^+	CH_3^+	NH^+	NH_4^+	HO^+	H_2O^+	H_3O^+
α/n in $10^{-7}cm^3s^{-1}$	4	4	4	1.7	6	1.3	6	3.7
ion	C_2^+	N_2^+	NO^+	O_2^+				
α/n in $10^{-7}cm^3s^{-1}$	9.4	5.7	3.7	3				

$$HC^+{=}O \; + \; e \rightarrow CH + O - 0.2 \text{ eV} . \tag{39}*$$

Turning to another aspect of dissociative recombination in which we are in conflict with other theorists we predict that the ejection of one hydrogen atom from a complex occurs much more readily than the ejection of two or more hydrogen atoms. To take a specific, but typical, example we predict that the branching ratio for

$$H_3O^+ + e \rightarrow H_2O + H \tag{40}$$

is almost unity and that the branching ratio for

$$H_3O^+ + e \rightarrow OH + 2H \text{ (or } H_2) \tag{41}$$

is small. The most obvious way for process (41) to occur is as a consequence of a radiationless transition involving molecular orbitals from the valence bonds between O^+ and two H atoms. Such a transition must be markedly less probable than the radiationless transition responsible for process (40): thus changes in more molecular orbitals are required and furthermore there is little overlap between the molecular orbitals of different valence bonds. There are two crossings (fig. 4). However this does not give rise to the product of two small Franck-Condon factors, it instead necessitates F_1 of eq. (34) being replaced by

$$F_2 = \int\int \chi_a(r)^2 \frac{dr}{d\varepsilon} \chi_b(s)^2 \frac{ds}{d\eta} \exp(-[\varepsilon+\eta]/kT)\, d\varepsilon d\eta \tag{42}$$

where ε and η are as indicated in fig. 4 and satisfy

$$\varepsilon + \eta = E \tag{43}$$

E being the energy of the free electron. Either ε or η may be negative so that the narrow range of integration which made F_1 small is not duplicated. Taking the two crossings to be identical we find from eq. (42) and (43) that

$$F_2 = F_1/\sqrt{2} \tag{44}$$

It might be suggested that more than 4.8 eV of the electronic energy released by the recombination (which is at most 6.4 eV) might appear as vibrational energy of H_2O so that dissociation into $O + H_2$ would ensue. Part of the energy converted comes from the action of the force between the separating H and H_2O fragments. However only 1/19 of this part can be given to the H_2O and it must be given mainly as translational energy of the molecule. The remainder comes from the difference between the

* The ionization potential of HCO is here taken from Rosenstock et al (1977). A lower value and therefore a greater endothermicity, is implicit in the data given by Lias et al (1984)

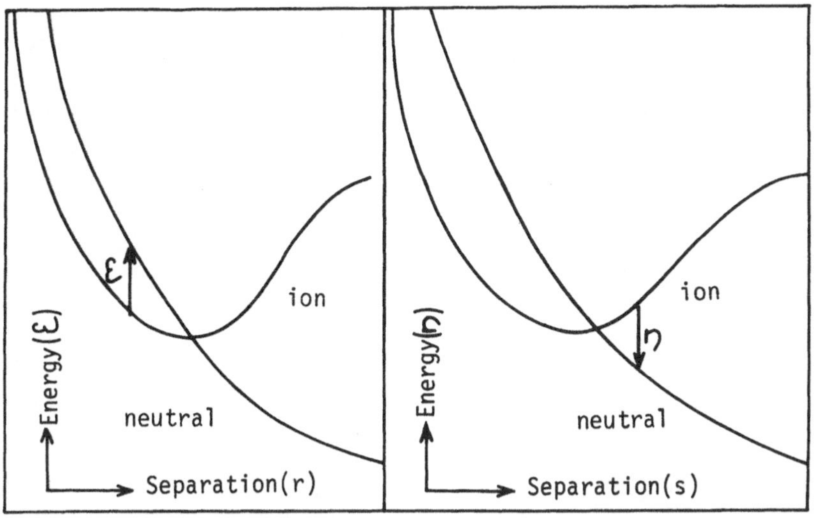

Fig. 4. Dissociative recombination with two crossings

equilibrium configuration of H_2O and that of three corresponding atoms of H_3O^+ and must be small because the ab initio quantal calculations of Lathan et al (1971) show virtually no difference in the OH bond lengths and a difference of only 14° in the HOH angle. The lowest electronically excited state of H_2O (cf Herzberg 1966) is somewhat out of reach and moreover has products of dissociation which are well out of reach. We conclude that channel (41) may be disregarded unless indeed the H_3O^+ has considerable vibrational energy initially (as it might have in some laboratory measurements).

When the conclusion just given was first presented (Bates 1986a) it was noted that Mitchell and McGowan (1985) had tentatively inferred that on the contrary channel (41) is dominant. Their inference was based on observations showing that H_2O is found only in small localized maser regions whereas H_3O^+ is widely dispersed throughout dense interstellar clouds. Lepp, Dalgarno and Sternberg (1987) have since suggested that H_2O is dissociated by photons produced by cosmic ray bombardment.

Ejection of H_2 as in processes (30) - (32) would be expected to be more improbable than the ejection of 2H because the formation of an H_2 bond further complicates the radiationless transition. The position is different if the structure of the polyatomic ion may be written $R^+.H_2$ where R^+ is an ion without a free valence and . denotes binding by the ion-induced dipole attraction. It is evident that if dissociative recombination of R^+ occurs the polarization attraction on H_2 would disappear.

The dissociative recombination of CH_5^+ has in the past been presumed to be a source of methane

$$CH_5^+ + e \rightarrow CH_4 + H \ . \tag{45}$$

However the ion has the structure $CH_3^+.H_2$ (Dyczmons, Staemmler and Kutzelnigg 1971) and the main dissociative channel must be

$$CH_3^+.H_2 + e \rightarrow CH_2 + H + H_2 \tag{46}$$

(Bates 1986a). Consistent with this picture the dissociative recombination coefficients of CH_5^+, CH_4^+ (the structure of which is $CH_3^+.H$) and of CH_3^+ are about the same. At 120K their values in units $10^{-7}\ cm^3\ s^{-1}$ are 11, 9.5 and 12 respectively (McGowan et al 1979). Having to abandon methane source (45) does not cause difficulty. Herbst (1987) has propounded an alternative source:

$$CH_5^+ + CO \rightarrow CH_4 + HCO^+ . \tag{47}$$

If the potential energy surfaces cross as required the polarization bond of $R^+.H_2$ could be dissolved by an excited state of R being formed in a single electron radiationless transition

$$R^+.H_2 + e \rightarrow R' + H_2 . \tag{48}$$

As suggested by Huntress and Mitchell (1979) conversion of R^+ into R can be brought about by process (48) preceded by

$$R^+ + H_2 \rightarrow R^+.H_2 + h\nu. \tag{49}$$

The efficiency of the sequence depends on the competition between channel (48) and other dissociative channels and is unknown.

Dissociative recombintion of $C_2H_3^+$ is another varient. The courses open to ions having the classical structure are

$$H_2C{=}C^+{-}H + e \rightarrow H_2C{=}C + H \tag{50}$$

$$\rightarrow H_2C + CH. \tag{51}$$

However as Herbst (1987) has noted it now seems (cf Raghavachari et al 1981) that the non-classical (bridged) structure has slightly lower energy. The hydrogen atom on the bridge is likely to be lost in any change so that the main dissociative recombination channels are

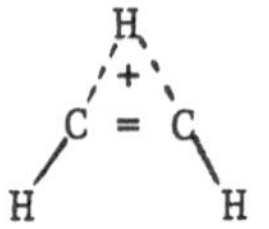

$$+ e \rightarrow C{\equiv}C{-}H + 2H \tag{52}$$

$$\rightarrow 2CH + H \tag{53}$$

Watson (1974, 1976) has drawn attention to the pecularity that although HNC lies 0.6 eV above HCN the abundances of the two isomers in dense interstellar clouds are seemingly around the same. He suggested that the explanation is that they are produced at approximately equal rates in the dissociative recombination of $HCNH^+$ which Pearson and Schaefer (1974) have shown is a linear ion. This would be in accord with the ideas on dissociative recombination advanced by Herbst (1978) and Green and Herbst (1979) - ideas which we do not accept.

The energy of a multiple bond between an atom and an ion is not independent of the other species bound to them and it is uncertain whether the structure of stable $HCNH^+$ is best written $H - C^+ = N - H$ (electronic wavefunction ϕ_1) or instead is best written $H - C \equiv N^+ - H$ (electronic wavefunction ϕ_2) the higher ionization potential of N as compared with C being compensated by the greater total valence bond energy. In the former case the main dissociation channels are

$$H - C^+{=}N{-}H + e \rightarrow H + CNH \tag{54}$$

$$\rightarrow CH + NH \tag{55}$$

while in the latter case they are

$$H - C{\equiv}N^{+}{-}H + e \rightarrow HCN + H \tag{56}$$

$$\rightarrow \; CN + NH \; . \tag{57}$$

The electronic wave function of the ion may be expressed approximately as the linear combination

$$\psi = a_1\phi_1 + a_2\phi_2 \; . \tag{58}$$

The relative importance of channels (54) and (56) depends inter alia on the ratio $|a_1|^2/|a_2|^2$ and is difficult to predict. Unless the energies of the two structures are almost identical the ratio is likely to be either small or large since ψ_1 and ψ_2 have little overlap in part of their multidimensional space. If dissociation channels (54) and (56) happen to be of about equal importance it is due to chance rather than the operation of a general rule. The problem we have just addressed has no relevance in connection with the classical and non-classical $C_2H_3^+$ structures where a linear combination of the electronic wavefunctions is not of interest because of the difference in the atomic configurations.

As an other example consider the factors which determine the efficiency of methanol production from CH_5O^+ by dissociative recombination. The principal limitation on the efficiency is the competition between the channels

$$CH_3{-}O^{+}H_2 + e \rightarrow CH_3{-}OH + H \tag{59}$$

$$\rightarrow CH_3 + H_2O \; . \tag{60}$$

The electronic wavefunction of CH_5O^+ may be expressed as a linear combination of amongst others, the electronic wavefunctions ϕ_1 and ϕ_2 of $CH_3 - O^{+}H_2$ and $CH_3^+.H_2O$ so that there is also some competition associated with

$$CH_3^+.H_2O + e \rightarrow CH_2 + H + H_2O \tag{61}$$

(Bates 1986a). Small overlap between ϕ_1 and ϕ_2 may make this competition minor.

To conclude this section I will list some dissociative recombination processes which have been mentioned in one of the following representative papers on interstellar cloud chemistry: Huntress and Mitchell 1979, Mitchell, Huntress and Prasad 1979, Herbst, Smith and Adams 1984. I will mark by an x those that I think should be discarded, and will give the likely competing channels (which may be uncertain because of uncertainty in the structure of the ion).

$$CH_2{=}CH{-}C{\equiv}N^{+}{-}H + e \rightarrow C \equiv C{-}C \equiv N + 2\,H_2 \qquad x \tag{62a}$$

$$\rightarrow CH \equiv C{-}C \equiv N + H_2 + H \qquad x \tag{62b}$$

$$\rightarrow CH_2{=}CH{-}C \equiv N + H \tag{62c}$$

$$\rightarrow CH_2{=}CH{-}C + NH \tag{62d}$$

$$CH_2{=}CH{-}C^{+}{=}N - H + e \rightarrow CH_2{=}CH + HCN \tag{62e}$$

$CH_2{=}C^+{-}C{=}O + e \rightarrow C_3O + H_2$ x (63a)

$\rightarrow CH_2 + C{=}C{=}O$ (63b)

$\rightarrow CH_2{=}C + CO$ (63c)

$CH_3{-}C^+{-}C{=}O + e \rightarrow C_3O + H_2 + H$ x (64a)

$\rightarrow CH_3 + C{=}C{=}O$ (64b)

$\rightarrow CH_3{-}C + CO$ (64c)

$CH_3{-}C^+{=}O + e \rightarrow CH_2{-}C{=}O + H$ x (65a)

$\rightarrow CH_3 + CO$ (65b)

(The 8.0 eV recombination energy is probably not enough for O to be freed).

$CH_3{-}N^+H_3 + e \rightarrow CH_2{=}NH + 3H$ x (66a)

$\rightarrow CH_3 + NH_3$ (66b)

$\rightarrow CH_3{-}NH_2 + H$ (66c)

$C_2H_5^+ + e \rightarrow CH \equiv CH + H_2 + H$ x (67a)

$CH_2 \overset{H}{\pm} CH_2 + e \rightarrow 2CH_2 + H$ (67b)

$\rightarrow CH_2{-}CH + 2H$ (67c)

$\rightarrow CH_2{-}CH_2 + H$ (67d)

$C_3H_5^+ + e \rightarrow CH_3{-}C \equiv C{-}H + H$ x (68a)

$CH_2{-}C^+H{-}CH_2 + e \rightarrow CH_2 + C_2H_3$ (68b)

$\rightarrow CH_2{=}C{=}CH_2 + H$ (68c)

$CH_3{-}C^+HOH + e \rightarrow CH_3{-}CHO + H$ x (69a)

$\rightarrow CH_3{-}CH + OH$ (69b)

$\rightarrow CH_3{-}COH + H$ (69c)

$\rightarrow CH_3 + CHOH$. (69d)

Observe that dissociative recombination can yield highly unsaturated components like C = C = O of (63b) or (64b) or metastable isomers like CH_3 - COH of channel (69c) or CHOH of channel (69d). The metastable isomers may in some instances change into the stable isomers e.g.

$CH_3{-}COH \rightarrow CH_3{-}CHO$, $CHOH \rightarrow HCHO$ (70)

The co-existence of HNC and HCN (see above) shows that the change does not invariably occur.

The association energy $E_o(I)$ of a complex ion I is the energy released when the ion is synthesised from its constituents S in the form of free neutral atoms and an atomic ion. For present purposes it is somewhat more convenient than the heat of formation $\Delta H(I)$ to which it is of course related by

$$E_o(I) = -\Delta H(I) + \sum_S \Delta H(S) \qquad (71)$$

where $\Delta H(S)$ is the heat required to produce atom or ion S from the element in its standard form (cf Benson 1976).

Information on $E_o(I)$ may be needed to determine whether an ion-molecule reaction is exothermic or endothermic. As will be seen in the next section it is needed in calculations on radiative association. It may enable us to identify the ionized atom in the polyatomic ion which is relevant in connection with the possible products of dissociative recombination. An extensive $E_o(I)$ or rather $\Delta H(I)$, compilation has been published (Lias Liebman and Levin 1984) but it does not cover all ions of interest.

In a classic paper Benson and Buss (1958) showed that valence bond energies in neutral molecules are quite accurately additive (usually within 0.1 eV). Table 2 contains some of their results (derived from the corresponding contributions to the heats of formation that were given originally). Those on = C = O and = C = C = are presented in a manner that acknowledges that the energies of double bonds depend on the species of atoms on the free valences.

It is plausible to assume additively of valence bond energies in an ionized molecule but there are also long-range interactions primarily ion-induced dipole, ion-permanent dipole and dipole-dipole interactions. Orientation dependent effects occur.

Seeking a simple approximate method of finding $E_o(I)$ we took the energy of a valence bond between a pair of neutral atoms to be as in table 2. To obtain the ion-induced dipole energies we used the known atomic polarizabilities (Miller and Bederson 1977) and estimated atomic separations but then allowed for the ability of the atomic ion concerned to accommodate positive charge (cf Taft 1979) by normalizing each set of ion-induced dipole

Table 2. Neutral-neutral bond energies corresponding to the contributions to the heats of formation recommended by Benson and Buss (1958).

Bond	Energy (k cal/mole)	Bond	Energy (k cal/mole)
C - H	98.7	O = C - H	$181.2 - \frac{1}{2} E_o(C = O)$
N - H	92.4	O = C - C	$172.4 - \frac{1}{2} E_o(C = O)$
O - H	108.9	O = C - O	$195.5 - \frac{1}{2} E_o(C = O)$
S - H	86.3		
C - C	82.7		
C - N	71.1		
C - O	84.5	C = C - H	$134.3 - \frac{1}{4} E_o(C = C)$
C - S	69.4	C = C - C	$121.5 - \frac{1}{4} E_o(C = C)$

Note: $E_o(C = O)$ and $E_o(C = C)$ represent the energies of the double bonds indicated.

Table 3. Ion-induced Dipole Energies

Ion	Intermediary Ion	Atom or group polarized	Energy (k cals/mole)
C^+	C	H	2.6
C^+	C	CH_3	5.7
C^+	O	H	4.5
C^+	O	CH_3	6.3
N^+	C	H	3.0
N^+	C	CH_3	6.0
O^+	C	H	7.1
O^+	C	CH_3	13.9

energies so as to reproduce the observed differences between I(E) for pairs such as

$$CH_3C^+HOH \quad \text{and} \quad C_2H_5C^+HOH \tag{72}$$

and such as

$$CH_3O^+H_2 \quad \text{and} \quad C_2H_5O^+H_2 \tag{73}$$

We then assigned to the energies of the valence bonds between ionized and neutral atoms the values (table 4) that best fit the $E_o(I)$ in the compilation of Lias et al (1984). The derived valence bond energies depend on the assumed ion-induced dipole energies and partly correct for the errors in these. The fit to $E_o(I)$ is usually to well within 0.2 eV but in some instances is only to within about 0.5 eV. Poor accord appears to be confined to ions like $CH_3C^+OH\,NH_2$ containing several different heavy neutral atoms and to ions in which there are multiple bonds.

Table 4. Ion Neutral Bond Energies (and some multiple bond energies).

Ion	Energy (k cals/mole)	Bond	Energy (k cals/mole)
$C^+ - H$	115.4	$C^+ = C$	173.0
$N^+ - H$	126.4	$C^+ = O$	223.2
$O^+ - H$	129.7		
$C^+ - C$	111.5	$C = C$	143.0
$N^+ - C$	101.1	$C \equiv C$	197.0
$C^+ - O$	134.1	$C = N$	141.0
$O^+ - C$	92.8	$C \equiv N$	208.5

Note: the multiple bond energies should be used with caution

Table 5. Effect of location of charge on bond energies in ions

Ion I	Charge on NH^+_{4-s}	$E_o(I)$ k cals/mole from bond energies	measured
NH_4^+	1.00	505.6	505.6
$CH_3-NH_3^+$	0.72	785.8	785.7
$(CH_3)_2-NH_2^+$	0.49	1066.0	1066.3
$(CH_3)_3-NH^+$	0.30	1346.2	1346.9

The results of ab initio quantal computations on polyatomic ions might seem to cast doubt on significance of ion-neutral bond energies. For example the computations of Ave et al (1976) on the $(CH_3)_s\ N^+H_{4-s}$ family show that the positive charge on N^+H_{4-s} resides mainly on the H atoms and that as s is increased from zero it decreases steadily by a total factor of over three (table 5). Nevertheless our $N^+ - H$ and $N^+ - C$ bond energies lead to association energies correct to 0.1 per cent (table 5). This remarkable accuracy is also attained in the $(CH_3)_s\ O^+H_{3-s}$ and $(CH_3)_s\ S^+H_{3-s}$ families (Bates 1987a).

The delocalization of the positive charge in a polyatomic ion means that the ionized atom is unique only with regard to its valency (which is different from that of the neutral atom). This raises the question of whether dissociative channels other than those proposed in the preceding section exist for instance the question whether

$$(CH_3)_3C^+ + e \rightarrow (CH_3)_2(CH_2)C + H \qquad (74)$$

may occur as well as

$$(CH_3)_3\ C^+ + e \rightarrow (CH_3)_2\ C + CH_3\ . \qquad (75)$$

In favour of a negative answer only the ionized atom has an orbital into which the free electron may enter.

RADIATIVE ASSOCIATION

The mean duration of a collision involving a molecule is longer than that of a collision between two atoms. Williams (1972) pointed out that this facilitates radiative association and may lead to rate coefficients much greater than those for processes (1). Black and Dalgarno (1973) thereupon recognized the likely importance of

$$C^+ + H_2 \rightarrow CH_2^+ + h\nu \qquad (76)$$

and Herbst and Klemperer (1973) initiated the inclusion of ion-molecule radiative association

$$A^+ + B \rightarrow (AB^+)^* \rightarrow AB^{\times} + h\nu \qquad (77)$$

in modelling work.

Discussions have recently been given (Herbst 1985a, Bates 1985) of the calculation of the rate coefficient k_r for radiative association and need not be repeated. We will here only cite some relevant formulae.

If A_r is the rate of stabilization of the activated complex $(AB)^*$ by photon emission the radiative association coefficient is

$$k_r = V(T)\, A_r \tag{78}$$

where the association volume V(T) is given by

$$V(T) = \frac{h^3}{(2\pi\mu kT)^{3/2}} \; \frac{\psi Q(AB^*)}{Q(A^+)A(B)} \tag{79}$$

in which μ is the A^+ - B reduced mass, ψ is the ratio of the electronic statistical weight of AB^* to the product of the electronic statistical weights of A^+ and B and the Q's are the internal partition functions indicated. The calculation of $Q(AB^*)$ includes the collisional aspects of the problem. It may be expressed

$$Q(AB^*) = \frac{1}{\sigma}\, \rho_{vib}\, Q_{rot}\, (AB^*) \tag{80}$$

where σ is the symmetry number of AB^*, ρ_{vib} is its energy density of vibrational states at the threshold energy and $Q_{rot}(AB^*)$ is in effect the contribution that rotation makes to the internal partition function. The expression for $Q_{rot}(AB^*)$ is cumbersome and will not be presented. It depends on the long-range A^+ - B interaction.

Let E_o and E_z be the association and zero point energies; let a be the Whitten-Rabinovitch factor; let the number of oscillators be increased by m to s when the complex is formed from the reactants; let ν_i be their frequencies and let ${}^rQ^I(rot)$ be the internal partition function for the r free internal rotations. Following Troe (1977) we then have that

$$\rho_{vib} = \left\{\frac{s-1}{s-1.5}\right\}^m \frac{(E_o + aE_Z)^{s-1}}{(s-1)\prod\limits_{i=1}^{s} h\nu} \; g^I(rot) \tag{81}$$

with

$$g^I(rot) = \frac{(s-1)!}{(s-\frac{1}{2}r-1)!} \left(\frac{(E_o + aE_Z)}{kT}\right)^{r/2} Q^I(rot)\ . \tag{82}$$

In the important case of a single free internal rotation of rotational constant B eq. (82) reduces to

$$g^I(rot) = \frac{(s-1)!}{(s-1.5)!} \left(\frac{\pi(E_o + aE_Z)}{B}\right)^{1/2} \tag{83}$$

Having cited the essential formulae we now turn to some calculations on k_r for

$$CH_3^+ + H_2 \rightarrow CH_5^+ + h\nu \tag{84}$$

This process is of considerable interest because of being a precursor to methane formation by process (47). Moreover it has been studied in the laboratory by Barlow, Dunn and Schauer (1984) using a trapped ion technique. It is necessary to treat para and ortho H_2 separately.

In the first of his recent papers on process (84) Bates (1985) obtained ρ_{vib} from formulae (81) taking E_o to be 1.7 eV, the then accepted value, and taking ν_i from ab initio quantal computations (Pople 1984, De Frees and McLean 1985). Making the dubious assumption that the barrier against the low frequency twist vibration becoming an internal rotation is high and noting that the ab initio quantal computations give that CH_5^+ has structure IV of Dyczmons, Staemmler and Kulzelnigg (1970) with the axis of H_2 perpendicular to the plane of a greatly distorted CH_3 and with none of the H atoms in equivalent positions so that $\sigma = 1$ he found from eq. (81) that

$$\frac{1}{\sigma}\rho_{vib} = 4.9 \times 10^2 (cm^{-1})^{-1} . \tag{85}$$

On combining this with the result of the trapped-ion measurements at 13K, which he accepted uncritically, he deduced that

$$A_r = 5.4 \times 10^2 \ s^{-1} . \tag{86}$$

The result is consistent with an estimate by Herbst (1985b) of the rate of radiative vibrational relaxation of the complex.

Some years earlier Smith, Adams and Algae (1982) had utilized their SIFT apparatus to make a primary set of measurements on the rate of

$$CH_3^+ + H_2 + He \rightarrow CH_5^+ + He \tag{87}$$

$$CH_2D^+ + H_2 + He \rightarrow CH_4D^+ + He \tag{88}$$

$$CHD_2^+ + H_2 + He \rightarrow CH_3D_2^+ + He \tag{89}$$

$$CD_3^+ + H_2 + He \rightarrow CH_2 D_3^+ + He \tag{90}$$

$$CD_3^+ + HD + He \rightarrow CHD_4^+ + He \tag{91}$$

and

$$CD_3^+ + D_2 + He \rightarrow CD_5^+ + He . \tag{92}$$

In principle their results should enable $\frac{1}{\sigma}\rho_{vib}$ to be determined more reliably than from the ab initio quantal computations (see below). However before they could be prudently used it was necessary to understand the striking variation in the measured rate coefficients down the family of processes. Having succeeded in reproducing the variation accurately by theory Bates (1986b) felt justified in exploiting the results. He took the efficiency with which the activated complex is stabilized in a collision with a He atom to be 0.3 as suggested by measurements of Cates and Bowers (1980) and deduced that

$$\frac{1}{\sigma}\rho_{vib} = 63 \ (cm^{-1})^{-1} . \tag{93}$$

The difference between values (85) and (93) is great. It is likely that the low frequency (110 cm^{-1}) twist vibration should be replaced by a free rotation when, as here, the complex carried the full association energy. In this circumstance the complex has the structure $CH_3^+.H_2$ and $\sigma = 6$ because the three H atoms of CH_3^+ and the two H atoms of H_2 are in equivalent positions. Taking E_o to be 1.91 eV (Herbst 1985a) and again adopting the frequencies obtained by the ab initio quantal computations we thence find from eq. (81) - (83) that

$$\frac{1}{\sigma}\rho_{vib} = 3.4 \times 10^2 \ (cm^{-1})^{-1} \tag{94}$$

which also differs greatly from value (93). The disagreement is scarcely a matter for concern because ν_i must be affected by the occurrence of free internal rotation: thus the CH_3^{+i} cannot remain planar.

The decrease in the adopted $\frac{1}{\sigma}\rho_{vib}$ value (85) to value (93) causes an increase in the A_r deduced from the result of the trapped ion measurement from value (86) to

$$A_r = 4.2 \times 10^3 \ s^{-1} \tag{95}$$

which exceeds the probable rate of radiative vibrational relaxation - calculations by Bates (1986c) give $9.2 \times 10^2 s^{-1}$ as an upper limit to this rate.

The difficulty just revealed naturally led to a re-examation of the trapped ion result. Barlow, Dunn and Schauer (1984) allowed circulating hydrogen gas at 13K to cool the initially very hot CH_3^+ ions (formed by CH_4 - He^+ collisions) that were confined in the trap before they started their measurements. From the noise power in image currents they non-destructively observed the rates of decrease of the total ion density and of $[CH_5^+]$; and they determined $[H_2]$ by monitoring the rate of cooling of ions which had been heated by a cyclotron pulse and applying a standard formula due to Cravath (1930). During the re-examination it was recognized (Bates 1986b) that this formula is inappropriate. It treats the collision between the ion and molecule as a collision between smooth elastic spheres which is an over-simplification. Account should be taken of the ion induced dipole interaction. When the CH_5^+ that is formed in a collision dissociates the direction of motion of either fragment in the centre of mass frame is random. Finally and most important the collision is not in general elastic because of the internal energy carried by the ortho hydrogen. According to Cravath's formula the mean fractional decrease in the kinetic energy of an ion in a collision with a molecule is

$$f = \frac{8m_1m_2}{3(m_1+m_2)^2} \quad \frac{T_1 - T_2}{T_1} \tag{96}$$

where m_1 and m_2 are the masses of the ionic and neutral species and T_1 and T_2 are their respective temperatures. The revised formula for this fraction is

$$f' = \frac{2m_1m_2}{(m_1+m_2)^2} \quad \frac{T_1 - T_2}{T_1} \quad - \quad \frac{2m_2 \overline{\Delta}}{3(m_1+m_2)kT_1} \tag{97}$$

$\overline{\Delta}$ being the average internal energy released in a collision.

Lengthy computations based on eq. (97) show that when the steady state is reached the CH_3^+ ions have a kinetic temperature of about 53K and moreover have considerable rotational energy. In consequence result (95) is changed to

$$A_r = 3.5 \times 10^4 \ s^{-1} \tag{98}$$

making the discrepancy with the estimated maximum rate of radiative vibrational relaxation worse, indeed making it so bad that the old tenet that radiative vibrational relaxation is responsible for the stabilization of the complex must be abandoned.

As an alternative Bates (1986c) proposed that the stabilization arises from the sequence of a nearly adiabatic electron transfer transition at an

Table 6. Inert Hydride Ions and $CH_4.H^+$

Ion	Polarizability of neutral* in A^3	Equilibrium neutral H+ distance* r_e in A	Dissociation energy† D in eV	$5\alpha r_e^{-2.6}$
$Ne.H^+$	0.40	0.99	2.08	2.1
$Ar.H^+$	1.64	1.28	4.17	4.3
$Kr.H^+$	2.48	1.50	4.35	4.3
$CH_4.H^+$	2.56			

* McDaniel and Mason (1973) † Huber and Herzberg (1979)

intersection of a specified pair of potential energy surfaces and a radiative transition between the two states concerned. It is convenient to let $\mathcal{S}(X^+,Y)$ denote both the surface and the state having X^+ and Y as dissociation products. According to Bates $\mathcal{S}(CH_3^+, H_2)$ and $\mathcal{S}(CH_3,H_2^+)$ are involved in CH_5^+ stabilization. It is possible that the transition between the $\mathcal{S}(CH_4^+, H)$ and $\mathcal{S}(CH_4,H^+)$ branches of the same pair of states contribute even though $\mathcal{S}(CH_4,H^+)$ lies 3.8 eV above $\mathcal{S}(CH_3^+,H_2)$ in the asymptotic region. Using the information in table 6 as his basis Bates argued that the $\mathcal{S}(CH_4,H^+)$ polarization well must be deep. The depth D for the inert hydride ions is approximately proportional to $\alpha r_e^{-2.6}$, α being the polarizability and r_e the equilibrium separation (table 6). If this empirical rule holds for $CH_4.H^+$ the depth of its well would be 5.7 eV, the maximum consistent with the observed proton affinity (Lias et al 1984), if the $C - H^+$ equilibrium distance were 1.36Å. It is of some relevance that the C - H equilibrium distance in methane is 1.08A. There seems no reason to suppose that the sequence proposed could not provide the requires stabilization rate of 3.5×10^4 s^{-1}.

Using a flowing afterglow Babcock and Streit (1984) have measured the rates of addition of F^- to BF_3 and of Cl^- to BF_3 and to BCl_3 in various ambient gases. They adduced evidence for a pressure independent process and suggested that the process is radiative association with A_r around 7×10^5 s^{-1} which is far too fast to be attributed to radiative vibrational relaxation. If their suggestion is correct it may plausibly be presumed that electron transfer related surfaces such as $\mathcal{S}(F^-,BF_3)$ and $\mathcal{S}(F,BF_3^-)$ are again involved. They lie close together in the asymptotic region, being there only 1.2 eV apart in the case of $\mathcal{S}(F^-,BF_3)$ and $\mathcal{S}(F,BF_3^-)$, which favours the occurrence of an intersection.

The failure of radiative vibrational relaxation to account for either the trapped ion measurements of Barlow, Dunn and Schauer (1984) on process (84) or the flowing afterglow results of Babcock and Streit (1984) on halide negative ion addition processes has an immediate repercussion.

It casts doubt on the perception that radiative vibrational relaxation governs the rate of the ion-molecule association processes of importance in interstellar clouds - aside from in process (76) which has long been regarded as exceptional (cf Herbst, Schubert and Certain 1977).

Considering process (77) let $\Delta E(A,B^+)$ denote the amount by which the potential surface on which A^+ and B approach one another, $\mathcal{S}(A^+,B)$, lies below the electron transfer related surface, $\mathcal{S}(A,B^+)$, in the asymptotic

Table 7. On stabilization by an electronic transition

Process identification number	Reactants	Well depths (eV) $W_o(A^+,B)$		$W_o(A,B^+) - \Delta E(A,B^+)$	
1	$CH_3^+ + NH_3$	a		4.6	E
2	$CH_3^+ + C_2H_4$	a		2.7	T
3	$CH_3^+ + CH_3OH$	a		3.6	E
4	$CH_3^+ + H_2CO$	a		2.9	T
5	$CH_3^+ + C_2H_2$	a		2.4	T
6	$CH_3^+ + H_2O$	a		2.9	E
7	$CH_3^+ + CO$	a		2.1	T
8	$H_2CN^+ + CO$	(6)	T	5	T
9 see text	$H_3O^+ + C_2H_4$	1.4	T	a-0.3,($\leq$1.4)	T
10 see text	$H_3O^+ + C_2H_2$	2.3	T	(2.3)	T

Note: a is the non-valence binding energy of the reactants due to the ion-dipole (induced and permanent) interaction; E and T indicate whether a value was obtained from experiment or from theory.

region. Let $W_o(A,B^+)$ be the amount by which the electronic energy of the $A - B^+$ lies below that of the sum of the electronic energies of A and B^+ the same atom being ionized in the complex as in the fragments. Taking into account the multi-dimensional nature of the potential energy surfaces it may be seen that the necessary and sufficient condition for an accessible intersection of the surfaces is that

$$W_o(A,B^+) > \Delta E(AB^+) \; . \tag{99}$$

Laboratory studies show that a nearly adiabatic transition between the surfaces usually occurs if there is such an intersection (cf McDaniel et al 1970).

The relative effectiveness of radiative stabilization by an electronic transition and by vibrational relaxation does not depend only on the transition probabilities. The depths of wells $\mathcal{S}(A^+,B)$ and $\mathcal{S}(A,B^+)$ measured with respect to the $(A^+ + B)$ limit are $W_o(A^+,B)$ say, and $[W_o(A,B^+) - \Delta E(A,B^+)]$ respectively. For example shoud one depth be much less than the other only a small fraction of the complex would be in the excited state which would greatly reduce the effective electronic transition probability. The circumstance would not affect vibrational relaxation.

Table 7 gives information relating to ten ion-molecule association processes identification number (T1) - (T10) that are of interest in interstellar cloud chemistry. All have intersecting electron transfer related potential energy surfaces and thus satisfy the prime criterion for stabilization by an electronic transition.

The depth $W_o(A,B^+)$ of the reactant well for processes (T1) - (T7) is determined by the ion-dipole interaction one or both of the reactants having no free valence. In at least some instants their values are quite likely to match $W_o(A,B^+) - \Delta E(A,B^+)$ closely enough to avoid prohibiting small Franck-Condon factors.

Processes (T9) and (T10) are affected by hydrogen atom migration. Jarrold et al (1986) have carried out important structure studies on $C_2H_7O^+$ and $C_2H_5O^+$ by collision-induced dissociation of a high energy ion beam. They found that with the reactants of (T9) and (T10) hydrogen atom migration occurs readily so that

$$H_3O^+ + C_2 H_4 \rightleftharpoons CH_3CH_2O^+H_2^* \qquad \text{(T9a)}$$

and

$$H_3O^+ + C_2H_2 \rightleftharpoons CH_3C^+HOH \quad . \qquad \text{(T10a)}$$

This gives rise to the possibility of stabilization by transitions between the electron transfer related states

$$\mathcal{S}(CH_3\ CH_2,\ H_2O^+) \text{ and } \mathcal{S}(CH_3C^+H_2,\ H_2O) \qquad \text{(T9b)}$$

$$\mathcal{S}(CH_3\ C^+H,\ OH) \quad \text{and} \quad \mathcal{S}(CH_3CH,O^+H) \quad . \qquad \text{(T10b)}$$

As far as can be judged the well depths are quite good matches (table 7). A study has not yet been made of whether hydrogen atom migration affects any of the other processes.

Electron transfer transitions are of the weak sub-Rydberg type (cf Herzberg 1950). Their weakness is due to the low overlap between the initial and final eigenfunctions. This overlap tends eventually rapidly towards zero as the distance between the nuclei on which the eigenfunctions are centred increases. For a given distance it must be greater for an ionized molecule than for a neutral molecule because of the strong polarization of the eigenfunctions by the free charge. Calculations on electron transfer transitions in polyatomic ions do not appear to have been done. However Arthurs, Bond and Hyslop (1957) have carried out calculations on the 1sσ- 2pσ, 2pσ-2sσ, 2pσ-3dσ and 2pσ-3dπ transitions of $He\,H^{2+}$ using the exact eigenfunctions. These transitions involve the transfer of an electron between the H(1s) orbital and a He^+(1s) or He^+(2s,p) orbitals. The He^{2+} perturbation is twice as strong as in the case in which we are interested but on the other hand the orbitals are much more tightly bound. Table 8 gives values of A_r obtained on taking the transition integrals to be as given by Arthurs, Bond and Hyslop and the photon energy of the radiation to be 0.5 eV. They suggest that it is not unreasonable to attribute the values of A_r given earlier, $3.5 \times 10^4\ s^{-1}$ and $7 \times 10^5\ s^{-1}$ to electron-transfer transitions but the crudeness of the model prevents a firm conclusion being reached.

Table 8. Electron Transfer Transition Probabilities

Transition (united atom notation)	He-H distance 3.5 a_o	4.5 a_o
	A_r s^{-1}	
$1s\sigma$ - $2p\sigma$	5.8×10^3	2.5×10^3
$1p\sigma$ - $2s\sigma$	3.7×10^3	1.5×10^3
$2p\sigma$ - $3d\sigma$	1.5×10^5	2.0×10^5
$2p\sigma$ - $2p\pi$	3.5×10^4	2.9×10^4

It is likely that stabilization by radiative electronic transitions of the electron transfer type makes some ion-molecule association rates faster than has been supposed thereby easing the synthesis of complex molecules in interstellar clouds.

Acknowledgement

I thank the United States Air Force for support under Grant No. AFOSR 850202.

REFERENCES

Adams N.G., Smith, D. and Clary, D.C., 1985, Rate coefficients for the reactions of ions with polar molecules at interstellar temperatures, Astrophys. J. (Letters), 296:L31.

Adams, W.S., 1941, Some results with the Coudé spectrograph of the Mount Wilson Observatory, Astrophys. J. 93:11.

Alge, E., Adams, N.G. and Smith, D., 1979, Measurements on the dissociative recombination coefficients of O_2^+ NO^+ and NH_4^+ in the temperature range 200-600K, J.Phys.B.At.Mol.Phys., 16:1433.

Arthurs, A.M., Bond, R.A.B. and Hyslop, J., 1957, The oscillator strengths of the $1s\sigma$ -$2p\sigma$, $2p\sigma$ -$2s\sigma$ and $2p\sigma$ - $2p\pi$ transitions of HeH^{2+}, Proc. Phys.Soc., 70A:617.

Ave, D.H., Webb, H.M. and Bowers, M.T., 1976, Quantitative proton affinities ionization potentials and hydrogen affinities of alkylamenes, J.Am.Chem.Soc., 98:311.

Babcock, L.M. and Streit, G.E., 1984, Third body effects in termolecular reactions. Halide ion addition to boron trifluoride and boron trichloride, J.Phys.Chem., 88:5025.

Bardsley, J.N., 1968, The theory of dissociative recombination, J.Phys.B. (Proc.Phys.Soc.), 1:365.

Barlow, S.E., Dunn, G.H. and Schauer, M., 1984, Radiative association of CH_3^+ and H_2 at 13K, Phys.Rev.Lett., 52:902 and 53:1610.

Bates, D.R., 1950, Dissociative recombination, Phys.Rev., 78:492.

Bates, D.R., 1982, Ion-polar molecule encounters, Proc.Roy.Soc.A., 384:289.

Bates, D.R., 1985, Radiative association of CH_3^+ and H_2, Astrophys.J., 298:382.

Bates, D.R., 1986a, Products of dissociative recombination of polyatomic ions, Astrophys.J.Lett., 30:L45.

Bates, D.R., 1986b, Some ter-molecular association processes in collisions of CH_3^+ and its deuterated analogs with H_2, HD and D_2, J.Chem.Phys., 85:2624.

Bates, D.R., 1986c, Interpretation of measured rate of radiative association of CH_3^+ and H_2, Phys.Rev. A34:1878.

Bates, D.R., 1987, Association energies of polyatomic ions, Int.J.Mass. Spectron. Ion Phys. (in preparation).

Bates, D.R. and Massey, H.S.W., 1947, The basic reactions in the upper atmosphere, Proc.Roy.Soc.A, 192:1.

Bates, D.R. and Mendas, I., 1985, Hitting collisions between ions and linear molecules having a quadrupole moment, Proc.Roy.Soc.A., 402:245.
Bates, D.R. and Morgan, W.L., 1987, Adiabatic invariance treatment of hitting collisions between ions and symmetrical top dipolar molecules, J.Chem. Phys.
Bates, D.R. and Spitzer, L., 1951, The density of molecules in interstellar space, Astrophys. J., 113:441.
Benson, S.W., 1976, Thermochemical Kinetics, p.288, Wiley New York.
Benson, S.W. and Buss, J.H., 1958, Additivity rules for the estimation of molecular properties: thermodynamic properties, J.Chem.Phys. 29:546.
Bhowmik, P.K. and Su, T., 1986, Trajectory calculations of ion-quadrupolar molecule collision rate coefficients, J.Chem.Phys., 84:1432.
Biondi, M.A. and Brown, S.C., 1949, Measurement of electron ion recombination, Phys.Rev., 76:1697.
Black, J.H. and Dalgarno, A., 1973, The formation of CH in interstellar clouds, Ap. Letters, 15:79.
Cates, R.D. and Bowers, M.T., 1980, Energy transfer in ion-molecule association. Dependence of collisional stabilization efficiency on the collision gas, J.Amer.Chem.Soc., 102:3994.
Clary, D.C., 1985, Calculations of rate constants for ion-molecule reactions using a combined capture and centrifugal sudden approximation, Molecular Phys., 54:605.
Clary, D.C., Smith, D. and Adams, N.G., 1985, Temperature dependence of rate coefficients for reactions of ions with dipolar molecules, Chem.Phys. Lett., 119:320.
Cravath, A.M., 1930, The rate at which ions lose energy in elastic collisions, Phys.Rev., 36:248.
Dalgarno, A., 1976, The interstellar molecules CH and CH^+ in Atomic Processes and Applications, P.G. Burke and B.L. Moiseiwitsch, eds, p.111, North Holland, Amsterdam.
Dalgarno, A., 1986, Is interstellar chemistry useful?, Quart.J.Royal Astron. Soc., 27:83.
De Frees, D.J. and McLean, A.D., 1985, Molecular orbital predictions of the vibrational frequencies of molecular ions, J.Chem.Phys., 82:333.
Douglas, A.E. and Herzberg, G., 1941, CH^+ in interstellar space and the laboratory, Astrophys.J., 94:381.
Duley, W.W. and Williams, D.A., 1984, Interstellar Chemistry, p.169, Academic, London.
Dyczmons, W., Staemmler, V. and Kutzelnigg, W., 1971, Near Hartree-Fock energy and equilibrium geometry of CH_5^+, Chemical Phys.Lett., 5:361.
Eyring, H., Hirschfelder, J.O. and Taylor, H.S., 1936, The theoretical treatment of chemical reactions produced by ionization processes, J.Chem.Phys., 4:479.
Gioumousis, G. and Stevenson, D.P., 1958, Reactions of gaseous molecular ions with gaseous molecules, J.Chem.Phys., 29:294.
Green, S. and Herbst, E., 1979, Metastable isomers: a new class of interstellar molecules, Astrophys. J., 229:121.
Herbst, E., 1978, What are the products of polyatomic ion-electron dissociative recombination reactions?, Astrophys. J., 222:508.
Herbst, E., 1985a, Radiative association rate coefficients under shocked conditions in interstellar clouds. The case of $CH_3^+ + H_2$, Astron. Astrophys., 153:151.
Herbst, E., 1985b, An update and suggested increase in calculated radiative association rate coefficients, Astrophys. J., 291:226.
Herbst, E., 1986a, An investigation of the effect of a centrifugal barrier on proposed large rate coefficients for ion-polar neutral reactions at low temperatures, Astrophys. J., 306:667.
Herbst, E., 1987, Gas phase chemical processes in molecular clouds in Interstellar Processes, Reidel, Dordrecht.
Herbst, E. and Klemperer, W., 1973, The formation and depletion of molecules in dense interstellar clouds, Astrophys. J., 185:505.

Herbst, E., Schubert, J.G. and Certain, P.R., 1977, The radiative association of CH_2^+, Astrophys. J., 213:696.
Herbst, E. and Leung Chun Ming, 1986, Effects of large rate coefficients for ion-polar neutral reactions on chemical models of dense interstellar clouds, Astrophys. J.
Herbst, E., Smith, D. and Adams, N.G., 1984, Ion-molecule synthesis of C_3O, Astron. Astrophys., 138:L13.
Herzberg, G., 1950, Molecular Spectra and Molecular Structure I Spectra of Diatomic Molecules, p.343 and 383, Van Nostrand, Toronto.
Herzberg, G., 1966, Molecular Structure and Molecular Spectra III Electronic Spectra and Electronic Structure of Polyatomic Molecules, p.342, Van Nostrand, Toronto.
Huber, K.P. and Herzberg, G., 1979, Molecular Spectra and Molecular Structure IV Constants of Diatomic Molecules, Van Nostrand Reinhold, New York.
Huntress, W.T. and Mitchell, G.F., 1979, The synthesis of complex molecules in interstellar clouds, Astrophys. J., 231:456.
Jarrold, M.F., Kirchner, N.J., Liv, S. and Bowers, M.T., 1986, J.Phys.Chem., 90:78.
Landau, L.D. and Lifshitz, E.M., 1960, Mechanics, §49, Pergamon, Oxford.
Lathan, W.A., Hehre, W.J., Curtiss, L.A. and Pople, J.A., 1971, Molecular orbital theory of the electronic structure of organic compounds, J.Amer.Chem.Soc., 92:6377.
Lepp, S., Dalgarno, A. and Sternberg, A., 1987, The cosmic ray ionization rate and the abundance of H_3^+ ions in dense interstellar clouds, Astrophys. J.
Lias, S., Liebman, J.F. and Levin, R.D., 1984, Gas phase basicities and proton affinities of molecules, J.Phys.Chem.Ref.Data, 13:695.
Mann, A.P.C. and Williams, D.A., 1980, A list of interstellar molecules, Nature, 283:721.
Marquette, J.B., Rowe, B.R., Dupeyrat, G., Poissant, G. and Rebrion, C., 1985, Ion polar molecule reactions: a CRESU study of He^+, C^+, N^+ + H_2O, NH_3 at 27, 68 and 163K, Chem.Phys.Lett., 122:431.
Massey, H.S.W., 1950, Negative ions (2nd Ed) p.50, University, Cambridge.
Massey, H.S.W. and Gilbody, H.B., 1974, Electronic and Ionic Impact Phenomena, Vol.IV, Recombination and Fast Collisions of Heavy Particles, p.2198, Clarendon, Oxford.
Millar, T.I., Adams, N.G., Smith, D. and Clary, D.C., 1985, The HCS^+/CS abundance ratio in interstellar clouds, Mon.Not.R.Astron.Soc., 216:1025.
Miller, T.M. and Bederson, B., 1977, Atomic and molecular polarizabilities - a review of recent advances, Adv.Atom.Molec.Phys., 13:1.
Mitchell, G.F., Huntress, W.T. and Prasad, S.S., 1979, Interstellar synthesis of the cyanopolyynes and related molecules, Astrophys.J., 233:102.
Mitchell, J.B.A. and McGowan, J.W., 1983, Experimental studies of electron-ion recombination in Physics of Ion-Ion and Electron-Ion Collisions, eds. F. Brouillard and J.W. McGowan, p.279, Plenum, New York.
Morgan, W.L. and Bates, D.R., 1987, Ion-dipolar molecule rate coefficients, Astrophys. J., 817:824.
McDaniel, E.W. and Mason, E.A., 1973, The Mobility and Diffusion of Ions in Gases, Wiley, New York.
McDaniel, E.W., Cermac, V., Dalgarno, A., Ferguson, E.E., Friedman, L., 1970, Ion Molecule Reactions p.321, Wiley, New York.
McGowan, J.W., Mul, P.M., D'Angelo, V.S., Mitchell, J.B.A., Defrance, P. and Froelich, H.R., 1979, Energy dependence of dissociative recombination below 0.08 eV measured with (Electron-ion) merged-beam technique, Phys.Rev.Lett., 42:81.
McKellar, A., 1940, Evidence for the molecular origin of some hitherto unidentified interstellar lines, Pub.Astron.Soc.Pacific, 52:187.
Pearson, P.K. and Schaefer, H.F., 1974, Some properties of H_2CN^+: A potentially important interstellar species, Astrophys. J., 192:33.
Pechukas, P. and Light, J.C., 1965, On detailed balance and statistical theories of chemical kinetics, J.Chem.Phys., 42:3281.

Pineau des Fôrets, G., Flower, D.R., Hartquist, T.W. and Dalgarno, A., 1986, Theoretical studies of interstellar shocks, Mon.Not.R.Astr.Soc., 220: 801.

Pople, J.A., 1984, Private communication.

Prasad, S.S. and Huntress, W.T., 1980, A model for gas phase chemistry in interstellar clouds, Astrophys. J. Supplement, 43:1.

Raghavachari, K., Whiteside, R.A., Pople, J.A. and Schleyer, P.R., 1981, Molecular orbital theory of the electronic structure of organic molecules, J.Am.Chem.Soc., 103:5649.

Rosenstock, H.M., Draxl, K., Steiner, B.W. and Herron, J.T., 1977, Energetics of gaseous ions, J.Phys.Chem.Ref.Data, 6, Suppl. 1.

Sakimoto, K., 1980, Ion-polar molecule reaction rates in interstellar clouds, Inst. Space Aeronaut. Sci., Tokyo Univ. Research Note, 102.

Sakimoto, K., 1981, Rotational excitation of symmetric top molecules by low energy ion impact, J.Phys.Soc. Japan, 50:1668.

Sakimoto, K., 1984, Orbiting collisions between ions and polar molecules: semi-classical PRS approaches, Chemical Phys., 85:273.

Smith, D., Adams, N.G. and Alge, E., 1982, Isotope exchange and collisional association in the reactions of CH_3^+ and its deuterated analogs with H_2, HD and D_2, J.Chem.Phys., 77:1261.

Su, T. and Bowers, M.I., 1975, Theory of ion polar molecule collisions, p.163 in Interactions between Ions and Molecules, P. Ausloos ed. Plenum, New York.

Su, T. and Bowers, M.I., 1979, Classical ion-molecule collision theory, p.84 in Gas Phase Ion Chemistry, M.T. Bowers ed. Academic, New York.

Swings, P. and Rosenfeld, L., 1937, Considerations regarding interstellar molecules, Astrophys. J., 86:483.

Taft, R.W., 1978, Proton transfer equilibria in the gas and solution phases, p.271 in Kinetics of Ion-Molecule Reactions, P. Ausloos ed. Plenum, New York.

Takayanagi, K., 1978, Low energy ion-polar molecule collisions - the perturbed rotational state approach, J.Phys.Soc.Japan, 45:976.

Takayanagi, K., 1979, Low energy ion-polar molecule collisions, Inst.Space Aeronaut. Sci. Tokyo Univ. Research Note 77.

Takayanagi, K., 1982a, Low velocity ion-molecule collisions with quadrupole interaction, Inst. Space Aeronaut. Sci., Tokyo Univ. Research Note 171.

Takayanagi, K., 1982b, Low velocity ion-molecule collisions with quadrupole interaction, J.Phys.Soc. Japan, 51:3337.

Takayanagi, K., 1982c, Low energy ion-molecule collisions, p.343 in Physics of Electronic and Atomic Collisions, S. Datz, ed. North Holland, Amsterdam.

Troe, J., 1977, Theory of thermal unimolecular reactions at low pressures, J.Chem.Phys., 66:4758.

Watson, W.D., 1974, Ion-molecule reactions, molecule formation and hydrogen-isotope exchange in dense interstellar clouds, Astrophys. J., 188:35.

Watson, W.F., 1976, Interstellar molecule reactions, Rev.Mod.Phys., 48:513.

Williams, D.A., 1972, Association reactions, Ap. Letters, 10:17.

Winans, J.G. and Stueckelberg, E.C.G., 1928, The origin of the continuous spectrum of the hydrogen molecule, Proc.Nat.Acad.Amer., 14:867.

PHOTO-IONISATION OF ATOMIC OXYGEN

M.J. Seaton

Department of Physics and Astronomy
University College London
Gower Street, London WC1E 6BT U.K.

The first calculation of the cross section for photo-ionisation of atomic oxygen was made by Bates, Buckingham, Massey and Unwin (1939) using a Hartree self-consistent field (scf) approximation, without exchange. The wave function Ψ_b for the initial bound state was taken to be the scf function for atomic oxygen calculated by Hartree and Black (1933) and the function for the final continuum state was

$$\Psi_c = \psi\theta \tag{1}$$

with ψ the scf function for O^+ and $\theta(\underset{\sim}{r})$ the wave-function for the ejected electron, calculated in the field of O^+. A calculation similar to that of Bates <u>et al</u>. was made by Yamanouchi and Kotani (1940).

For atomic oxygen in its ground configuration, near-threshold photo-ionisation can occur with ejection of an s electron,

$$2p^4 + h\nu \rightarrow 2p^3ks \tag{2}$$

or ejection of a d electron,

$$2p^4 + h\nu \rightarrow 2p^3kd, \tag{3}$$

$k = mv/\hbar$ being the wave-number for the ejected electron. Bates <u>et al.</u> found that:

(i) the cross-section for ejection of a d electron is larger than that for ejection of an s electron;

(ii) the wave-function for a d electron in the field of O^+ is not very different from a Coulomb d-wave function.

Similar results could be expected for other atoms with outer 2p electrons. Bates (1939) calculated photo-ionisation cross-sections for all the elements from boron to neon using the ks and kd functions calculated in the field of O^+, the justification for this procedure being: that the ejection of s electrons gives only a small contribution to the total cross-sections, and hence the use of an approximate ks function should not lead to large errors; and that the kd functions calculated in the field of the other ions would be similar to those calculated in the field of O^+. Later, Bates (1946a) gave a general formula for the calculation of photo-ionisation cross-sections in the approximation of using Slater-type orbitals for bound electrons and Coulomb functions for ejected electrons.

During the years 1939 to 1945 research in atomic physics was slowed down but by no means halted. In 1946 Bates (1946b) published a general survey of atomic photo-ionisation in which he wrote that this was "rendered possible by the insight gained from the detailed calculations which have been carried out in recent years". In that review he was already able to explain many of the general trends in the behaviours of the cross-sections as functions of frequency ν, and in particular why the cross-sections for complex ions differ from those for atomic hydrogen, which decrease approximately as ν^{-3}. He showed that these trends could be understood on making a systematic study of interference effects in the calculation of the dipole matrix elements, and the way in which these effects depend on the energy of the ejected electron. He was thus able to explain the increase, with increasing ν, of the near-threshold cross-section for neon and the minima in the cross-sections for the alkali atoms, which are due to interference effects giving changes in the signs of the matrix elements. Some of these explanations were re-discovered many years later.

The first calculation of a photo-ionisation cross-section allowing for exchange was made by Bates and Massey (1941) who considered the very difficult case of Ca. For the initial state they used a function Ψ_b calculated from solutions of the Hartree-Fock equations (scf with exchange) and for the final state they took

$$\Psi_c = A\,\psi\theta \tag{4}$$

where ψ is a Hartree-Fock function for Ca^+, $\theta(\sigma,\underset{\sim}{r})$ (σ being a spin co-ordinate) a function for the ejected electron, and A an anti-symmetrisation operator. The function θ was obtained on solving an integro-differential equation. I have discussed elsewhere (Seaton, 1982) later work on the problem of Ca photo-ionisation.

By 1946 there was a great deal of interest in the calculation of the cross section for photo-detachment from H^-, due to the importance of this process in stellar atmospheres. The first H^- calculation was made by Bates and Massey (1940) and subsequent very elaborate calculations were made by Chandrasekhar and his collaborators who used both the dipole-length and dipole-velocity operators and found that, in order to obtain accurate results, it was essential to use an elaborate wave-function for the initial bound state. A comprehensive review of work on H^- photo-detachment has been published by Bates (1978). A special feature of the problem is that one can use wave-functions which depend explicitly on the distance between the two electrons, r_{12}. For heavier atoms, such as oxygen, the nature of the problem is somewhat different.

The process of photo-ionisation of atomic oxygen is of importance for the formation of the F layers in the earth's ionosphere. Bates and Seaton (1949) calculated the threshold cross-section allowing for exchange and using both the length and velocity formulations. For the initial state they used the Hartree-Fock function of Hartree, Hartree and Swirles (1939) and in solving the integro-differential equations for the ejected electron both authors did comparable amounts of work in turning the handles of their calculating machines. The cross-section as a function of energy was obtained on normalising the results from the general formula of Bates (1946a) to the more accurate threshold calculations, and similar but slightly less accurate calculations were made for carbon and nitrogen. The results of Bates and Seaton for the cross-sections as functions of energy are shown on Figure 1 and results of various calculations for the threshold cross-sections are given in Table 1. In addition to the points (i) and (ii) which had emerged from the earlier work, a third point emerged from the 1949 calculations:

(iii) the near-threshold cross-section is sensitive to the wave function used for the initial bound state.

The reason for this sensitivity is that the wave-function for the ejected d-electron is small for small values of r, due to centrifugal repulsion, and the radiative matrix element is therefore sensitive to the outer part of the wave-function for the bound electron. This sensitivity is greatest using the length formulation. The dipole-length threshold cross-section was found to be $\sigma = 4.4$ Mb without exchange and 2.3 Mb with exchange (1 Mb = 10^{-18} cm^2). The difference between these two numbers is due almost entirely to the differences between the 2p radial functions, with and without exchange.

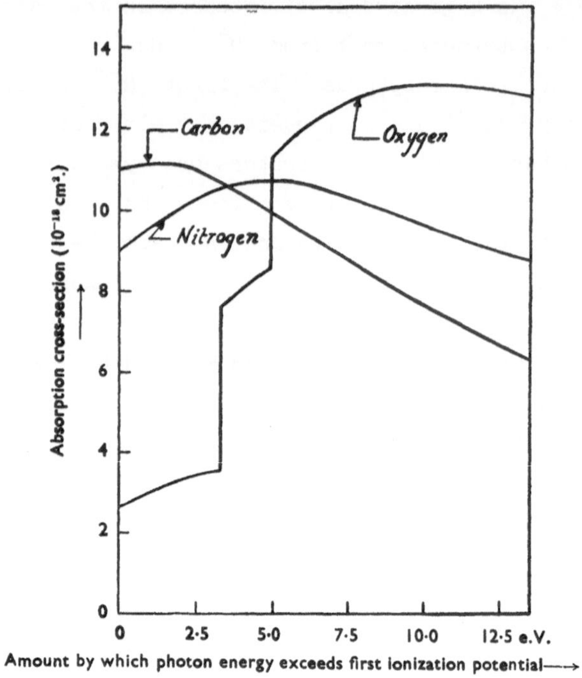

Absorption cross-section curve for atomic oxygen, nitrogen and carbon.
(Note: The breaks in the oxygen curve are at the second and third ionization potentials.)

Figure 1. Photo-ionisation cross-sections for C, N and O. From Bates and Seaton (1949).

Calculations for higher energies were made by Dalgarno and Parkinson (1960) who included contributions from the process

$$2s^2 2p^4 + h\nu \rightarrow 2s2p^4 kp \tag{5}$$

for ejection of a 2s electron in addition to the processes (2) and (3) for ejection of a 2p electron. They used Coulomb functions for the ejected electron and the dipole-velocity formula which is considered to be more accurate than the length formula at high energies (Dalgarno and Lewis, 1956). Further progress in the calculation of the atomic oxygen photo-ionisation cross-section has depended on the use of electronic computers. I think it likely, but cannot prove it, that the total amount of computation which has by now been devoted to this problem is larger than the total amount of scientific computation done by all mankind prior to, say, 1939. Calculations at wavelengths from the threshold at 910 Å out to 25 Å were made by Dalgarno, Henry and Stewart (1964) using a computer at the NASA Goddard Space Flight Center in New York, all three authors being at that time on leave from The Queen's University of Belfast. The basic Hartree-Fock formulation used was similar to that of Bates and Seaton (1949). The

Table 1. Calculated values of the threshold cross-section, in Mb, for photo-ionisation of atomic oxygen (L for dipole length, V for dipole velocity).

Reference	Method	σ L	σ V
Bates, 1939	Hartree self-consistent field	4.4	3.6[†]
Bates and Seaton, 1949	} Hartree-Fock	2.3	2.8
Dalgarno et al., 1964		2.7	3.4
Henry, 1967		2.2	2.7
Starace et al., 1974		2.5	2.9
McGuire, 1968	Herman-Skillman	2.7	—
Thomas and Helliwell, 1970	Klein-Brueckner	2.1	—
Kähler, 1971	Scaled Thomas-Fermi } central potential	3.0	—
Ganas, 1973	2-parameters	5.5	—
Starace et al., 1974	Herman-Skillman	2.8	—
Henry, 1968	Hartree-Fock for Ψ_b, close-coupling for Ψ_c	2.6	2.9
Taylor and Burke, 1976	} Close-coupling for Ψ_b and Ψ_c	4.1	—
Pradhan and Saraph, 1977		4.0	—

[†] Value given by Bates and Seaton, 1949.

cross-sections were calculated for photo-ionisation of atomic oxygen in its ground state $2s^2 2p^4\ {}^3P$, and production of O^+ in the states $2s^2 2p^3\ {}^4S^o$, ${}^2D^o$ and ${}^2P^o$ and $2s2p^4\ {}^4P$ and ${}^2P$. Tables 1 includes results from four separate calculations of the threshold cross-section all in the same Hartree-Fock approximation and they are seen to be in tolerably good agreement. The Table also includes results from four calculations in which central potentials are used to calculate the wave-functions for both the initial and the final states (some of these calculations were made in the course of systematic studies of photo-ionisation for many atomic systems): there is seen to be considerable scatter in the results obtained.

At energies in the ionisation region one can detect absorption of photons, and hence measure the photo-absorption cross-section, or detect product ions or ejected electrons and measure the photo-ionisation cross-section. The first measurement of the photo-absorption cross-section of atomic oxygen was made by Cairns and Samson (1965) who used atomic oxygen from an O_2-He microwave discharge and monochromatic photon fluxes from a light source giving strong emission lines. The measured photo-absorption cross-section, which was estimated to have an accuracy of ± 30 per cent, is shown on Figure 2 and compared with the photo-ionisation cross-section of Dalgarno et al. (1964). The measured cross-section is seen to be systematically larger than the calculated one and to be very much larger at four of the wavelengths used in the experimental work, 780, 736, 725 and 686 Å. Cairns and Samson interpreted the additional absorption at

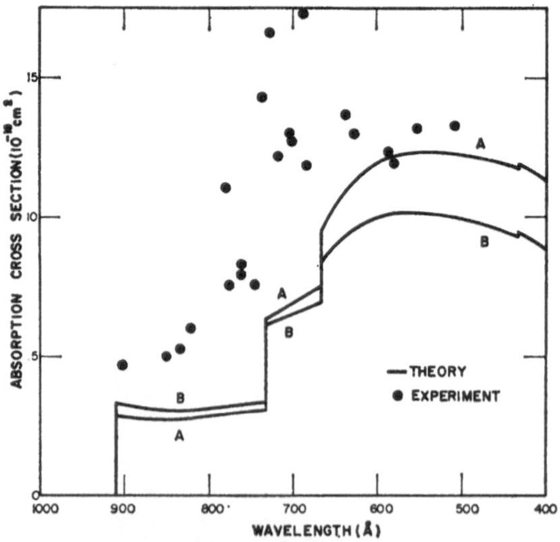

Figure 2. The atomic oxygen photo-absorption cross-section from Cairns and Samson (1965), and the photo-ionisation cross-section from Dalgarno et al. (1964) (curves A for dipole length, B for dipole velocity).

these wavelengths as being due to spectrum lines. They were careful to distinguish between photo-absorption and photo-ionisation, noting that "if an atom is excited to a discrete state lying above its lowest ionisation potential, it can de-excite with the emission of radiation or, selection rules permitting, can undergo a radiationless transition to an adjoining continuum". If this auto-ionisation transition does not occur the cross-section for photo-absorption can be larger than that for photo-ionisation. Cairns and Samson noted that if, on the other hand, auto-ionisation does occur then at the wavelengths affected the photo-ionisation cross-section could be much larger than the values which had been calculated up to that time.

A year after the publication of the work of Cairns and Samson, Huffman et al. (1966) published a detailed spectroscopic study of absorption by atomic oxygen in the ultra-violet. A continuum light-source was used and wavelengths of absorption features were measured to an accuracy of 0.02 Å (term values correct to about 2 cm^{-1}). The spectra were recorded on photographic plates and no attempts were made to measure cross-sections. Some of the observed lines were found to be diffuse, indicating broadening by auto-ionisation. For transitions from the $2p^4\ {}^3P$ initial state the final states allowed in LS coupling have angular momenta and parity of $SL\pi = {}^3S^o,\ {}^3P^o$ and ${}^3D^o$. Most of the observed features in the UV spectra were identified with transitions to discrete states of the type

$$2s^2 2p^3 (S_i\ L_i\ \pi_i) n\ell \quad SL\pi\ . \tag{6}$$

The discrete states occur at energies E which are smaller than the energy $E(S_i\ L_i\ \pi_i)$ for the O^+ parent state. For $E > E(S_i\ L_i\ \pi_i)$ we have continuum states

$$2s^2 2p^3 (S_i\ L_i\ \pi_i) k\ell \quad SL\pi. \tag{7}$$

The allowed values of ℓ for different parent states $S_i\ L_i\ \pi_i$ and different values of $SL\pi$ in equations (6) and (7) are given in Table 2.

Table 2. Values of ℓ for states $2s^2 2p^2 (S'L'\pi')\ell\ SL\pi$ having optically-allowed transitions to $2s^2 2p^2\ {}^3P$

S'L'π' SLπ	${}^4S^o$	${}^2D^o$	${}^2P^o$
${}^3S^o$	s	d	-
${}^3P^o$	-	d	s,d
${}^3D^o$	d	s,d	d

Let us consider the case of energies E in the range $E(^4S^o) < E < E(^2D^o)$. For $SL\pi = {}^3S^o$ we have continuum states $(^4S^o)ks$ and discrete states $(^2D^o)nd$ and it is the interaction between states of these two types which leads to auto-ionisation. Continuing to consider the range $E(^4S^o) < E < E(^2D^o)$, the position for $^3P^o$ states is different in that one can have discrete states $(^2D^o)nd$ and $(^2P^o)ns$ and nd but there are no continuum $^3P^o$ states in this energy range. It follows that the $^3P^o$ states cannot auto-ionise in LS coupling. Auto-ionisation can occur when one allows for departures from LS coupling which result from inclusion of relativistic effects.

A comprehensive compilation of O I energy levels, including those for many auto-ionising states, is given by Moore (1976).

Absolute measurements of the atomic oxygen photo-ionisation cross-section were obtained by Comes et al. (1968) who used an emission-line light-source and detected product ions in a mass spectrometer. This work showed conclusively that the ionisation cross-section contains a number of peaks due to auto-ionisation, at wavelengths in agreement with those obtained in the absorption work. The peak closest to the ionisation threshold was at a wavelength of 879 Å and could be identified with a transition to the state $2p^3(^2P^o)3s\ {}^3P^o$ which does not auto-ionise in LS coupling; the experiment showed that auto-ionisation of this state does, nevertheless, occur. The absolute cross-section of Comes et al. was found to be smaller than that of Cairns and Samson and in better agreement with the best estimates which had, at that time, been obtained from theory.

Further measurements for photo-ionisation were made by Dehmer et al. (1973) using a continuum light-source and detection of product ions. This work gave results in the range 920 to 650 Å at a resolution of 0.42 Å and, for the seven lowest auto-ionising multiplets, at a resolution of 0.16 Å. Figure 3 shows the photo-ionisation yield in the near-threshold region, with the thresholds for the fine-structure components of the initial state, $2p^4\ {}^3P_J$, resolved. On making the reasonable assumption that the cross-section σ is the same for each of the J levels, Dehmer et al. calculated the relative populations of these levels and found that they could be fitted to a Boltzmann distribution at $T = 375 \pm 25$ K. All of the main features which had been seen in absorption were detected in the ionisation experiment of Dehmer et al., and hence shown to be auto-ionising. For many of the multiplets the fine-structure was resolved, or at least partially resolved, but the resolution was not sufficient for the measurement of auto-ionisation widths. Dehmer et al. compared the

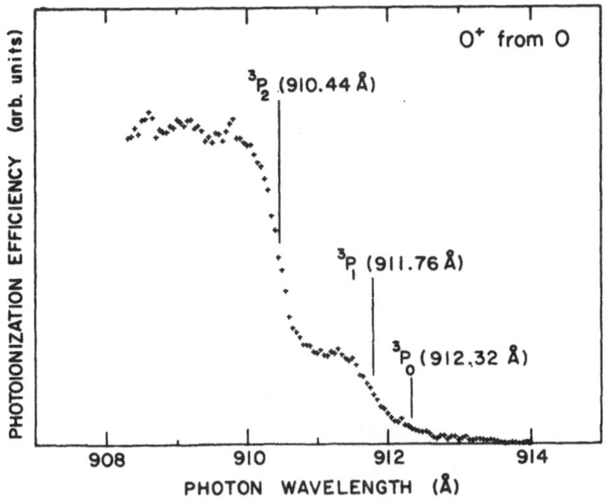

Figure 3. The photo-ionisation efficiency of atomic oxygen in the vicinity of the first ionisation threshold. From Dehmer et al. (1973).

measured relative strengths of components of multiplets with values calculated using the measured distribution among the initial J levels and assuming that: (i) the relative photo-excitation rates to the final fine-structure levels are given by the usual LS-coupling formulae; and (ii) that all photo-excitations are followed by auto-ionisation. Good agreement was obtained for all of the terms except $2s^2 2p^3 (^2P^o) 3s \; ^3P^o$ and $2s2p^5 \; ^3P^o$, which do not auto-ionise in LS coupling. The discrepancies for these two terms must be due to at least one of the two assumptions made being incorrect.

More recent measurements of the absolute cross-sections have been made by Kohl et al. (1978) and by Samson and Pareek (1985). Kohl et al. worked at 5 wavelengths in the near-threshold region and used a shock-heated gas as a source of atomic oxygen. The experimental technique used in the atomic oxygen work was also used to measure the photo-ionisation cross-section of atomic hydrogen and the results obtained were in close agreement with those from accurate quantum-mechanical calculations. The results for the near-threshold atomic oxygen photo-ionisation cross-section were estimated to be correct to within ± 7 per cent. Samson and Pareek (1985) made relative measurements in the extended range from 900 to 120 Å and normalised to absolute measurements at the wavelengths of He I λ584 and He II λ304. Below the $^2P^o$ threshold they used wavelengths avoiding known auto-ionisation peaks but above that threshold they took data at all available wavelengths, since no previous studies had been made of auto-ionisation structures in that region. Auto-ionisation features identified with $2s2p^4(^4P)3p$ were detected. The absolute cross-section in the range 900 to 120 Å was estimated to be correct to within ± 9 per cent.

All of the experimental results, and also the theoretical results which I shall discuss later, show that apart from the narrow $2p^3(^2P^o)3s\ ^3P^o$ feature near 879 Å the atomic oxygen photo-ionisation cross-section does not contain any resonances in the range of $\lambda > 820$ Å and that in this range the cross-section has little variation as a function of wavelength. The means of measurements made in this region may therefore be used to obtain estimates of the near-threshold cross-section. The results are shown in Table 3. The two most recent measurements, those of Kohl *et al*. and of Samson and Pareek, are in good agreement and give a mean value of

$$\sigma = 4.3 \pm 0.3 \text{ Mb}. \tag{8}$$

This is smaller than the cross-section of Cairns and Samson, although within the error estimate, but definitely larger than the value obtained by Comes *et al*.

Table 3. Measured values of the threshold cross-section, in Mb, for photo-ionisation of atomic oxygen.

Reference	σ
Cairns and Samson, 1965	5.0 ± 1.4
Comes *et al*., 1968	2.3 ± 0.4
Kohl *et al*., 1977	4.6 ± 0.3
Samson and Pareek, 1985	4.0 ± 0.4

Having described some of the experimental work on atomic oxygen photo-ionisation carried out over the past 21 years, I return to a discussion of theoretical work on the problem during the same period. The experimental work posed two problems for the theoreticians: to make calculations allowing for auto-ionisation; and to improve the accuracy of the calculations (on this second point the situation was at first confused since different experiments had given discrepant values for the absolute cross-sections). Advances in the theory depended on abandoning the idea that the final state could be described in terms of an ejected electron moving in the field of just one state of a product ion. An improved theory is obtained on replacing (4) by the expansion

$$\Psi_c = \mathcal{A} \sum_i \psi_i \theta_i \tag{9}$$

where ψ_i is a wavefunction for the product ion in state i. Use of (9) with optimised functions θ_i is known as the close-coupling (CC) approximation. The total energy is

$$E = E_i + \frac{1}{2} mv_i^2 \qquad (10)$$

where E_i is the ion energy and $\frac{1}{2} mv_i^2$ the kinetic energy of the ejected electron. We refer to i as the channel index. The electron wave-number for channel i is $k_i = mv_i/\hbar$. For $k_i^2 > 0$ channel i is said to be open and gives continuum states of the type (7) while for $k_i^2 < 0$ channel i is said to be closed. One closed channel, in the absence of open channels, gives discrete bound states of the type (6). Auto-ionisation arises from the coupling between open and closed channels.

The need to consider expansions of type (9) had arisen in connection with the problem of calculating cross-sections for excitation of atoms and ions by electron impact, a process which depends on the couplings between open channels. Electron-impact excitation of the metastable $2s^2 2p^3\ {}^2D^o$ and ${}^2P^o$ levels at O^+ is of interest because the transitions in "forbidden" lines, ${}^2D^o \rightarrow {}^4S$, λλ3726, 3729 and ${}^2P^o \rightarrow {}^2D^o$ λλ7320, 7330 are observed in the spectra of gaseous nebulae. Given the cross-sections for excitation and de-excitation of these levels, the probabilities for the radiative transitions, and the observed strengths of the spectrum lines one can obtain important information concerning the temperatures and densities in the nebulae and the abundances of oxygen ions.

Optimisation of the functions θ_i in (9) gives a system of coupled integro-differential equations (Seaton, 1953a) and the first approximate solutions of these equations for the $(e + O^+)$ system were obtained by Seaton (1953b), using hand computing. In that and subsequent work on the problem it was found that the strongest coupling occurred between channels of the type

$$2p^3(S_i\ L_i\ \pi_i)k_i p\ {}^3P \text{ and } {}^3D \qquad (11)$$

having even parity. For the odd-parity final states of the photo-ionisation problem the coupling was found to be much weaker and it is this which causes the auto-ionisation widths to be narrow.

The first calculations for O photo-ionisation using a CC expansion for the final state were made by Henry (1967, 1968), who continued to use a Hartree-Fock function for the initial state. The 1967 calculations were made with inclusion of open channels only in the CC expansion, and hence did not give auto-ionisation resonances, but the closed channels were included in the 1968 work. The auto-ionisation probabilities Γ

were expressed in wavelength units and found to be small: below the $^2D^o$ threshold the largest value obtained was $\Gamma = 0.0203$ Å for the state $(^2D^o)3d\ ^3D^o$ and above that limit the largest value was $\Gamma = 0.134$ Å for $(^2P^o)4s\ ^3P^o$. These are much smaller than the resolution widths in the experiment of Dehmer et al. (1973) and one can therefore understand why the observed line-profiles did not differ from what would be expected considering instrumental broadening alone. Calculations in an approximation similar to that of Henry were described by Rudd and K. Smith (1968) who compared the positions of the resonances with positions obtained experimentally in proton-bombardment of atomic oxygen.

Apart from work on angular distributions, which I shall discuss later, no significant improvements in the theory were made before the 1976 work of Taylor and Burke (further details are given by Taylor, 1976) and similar work of Pradhan and Saraph (1977: further details are given by Pradhan, 1976 and 1978). The new feature introduced was the use of CC expansions for both the initial and the final state. This gives an improved initial-state wave-function and a more balanced approximation. In practice the functions θ_i are constrained to be orthogonal to the orbital functions used for the ion and, to compensate for this constraint, additional terms are included in the expansion to give

$$\Psi = A \sum_i \psi_i\, \theta_i + \sum_j \Phi_j\, c_j \tag{12}$$

where the Φ_j functions are of bound-state type for the whole system. Further functions Φ_j can be included to improve the accuracy. The use of CC expansions to obtain wave-functions for bound states is sometimes referred to as the frozen-cores approximation.

The work of Taylor and Burke on the one hand, and of Pradhan and Saraph on the other, used similar formulations of the problem but completely different numerical methods. In both cases the ion states

$$2s^2 2p^3\ ^4S^o,\ ^2D^o,\ ^2P^o \quad \text{and} \quad 2s2p^4\ ^4P,\ ^2D,\ ^2S,\ ^2P \tag{13}$$

were included, and Taylor and Burke also included $2p^5\ ^2P^o$. In addition to the basic spectroscopic configurations, Pradhan and Saraph included the correlation configurations $2s2p^3\ \bar{3}d$ and $2s^2 2p^2\ \bar{3}d$. Numerical solutions were obtained by Taylor and Burke using the R-matrix method (Burke et al., 1971) and by Pradhan and Saraph using the method of reduction to linear algebraic equations (Seaton, 1974). The results obtained were in close agreement. The elaborate frozen-cores function used for the initial bound state should be much more accurate than the Hartree-Fock function used in much of the earlier work. The use of this elaborate function gives a threshold cross-section of 4.0 Mb compared with the best estimate from experiments of 4.3 ± 0.3 Mb and the Hartree-Fock values of 2.3 Mb.

Figure 4 compares the results of the calculations by Taylor and Burke, and by Pradhan and Saraph, with the experimental results of Samson and Pareek. At energies below the $^2P^o$ threshold most of the results shown on this figure are for the non-resonant background cross-section. The figure also includes the cross-section calculated by Starace et al. (1974) using wave-functions and potentials from Herman and Skillman (1963). Samson and Pareek comment that this calculation, which does not allow for any correlation effects, gives an agreement with the experimental results which is extraordinary. Perhaps even more extraordinary is that, in the near-threshold region, even better agreement with experiment is given by the original 1939 calculation of Bates! It would be unwise to make the generalisation that simple models are always likely to give better results than the most elaborate modern calculations.

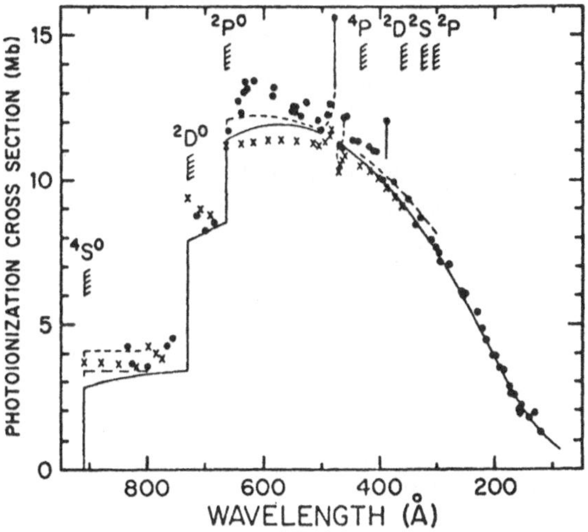

Figure 4. The atomic oxygen cross-section, from Samson and Pareek (1985). The experimental values of Samson and Pareek are shown by filled circles. Calculated values are: full line, Herman-Skillman results of Starace et al. (1974); crosses from Pradhan and Saraph (1977); dashed lines from Taylor and Burke (1976) (short dash for dipole length, long dash for dipole velocity).

Pradhan and Saraph made calculations for both bound-bound and bound-free transitions. The absorption cross-section for a bound-bound transition, $b \to a$, is

$$\sigma_\nu(b \to a) = \frac{\pi e^2}{mc} f(a,b)\phi_\nu \qquad (14)$$

where ϕ_ν is a normalised profile factor, $\int \phi_\nu d\nu = 1$, and $f(a,b)$ the dimensionless oscillator strength

$$f(a,b) = \frac{(E_a - E_b)}{E_R} |D|^2/(3g_b). \qquad (15)$$

In (14), g_b is the statistical weight of the initial level, E_a and E_b the energies of the initial and final levels, E_R the Rydberg unit of energy ($E_R = me^4/(2h^2) \simeq 13.6$ eV) and D is the reduced dipole matrix element,

$$D = (\Psi_a||\vec{R}||\Psi_b) \qquad (16)$$

where $\vec{R} = \sum_m \vec{r}_m$ the summation being over all of the electrons in the system. The matrix element (16) is evaluated in atomic units. A relation between bound-bound and bound-free absorption is obtained on considering final states a just below an ionisation limit with energy E_i. For such states we have

$$E_a = E_i - E_R/(n - \mu)^2 \qquad (17)$$

where n is the principal quantum number and μ the quantum defect. Let ΔE_a be the separation between two adjacent levels, with quantum numbers n and $(n + 1)$. For n large, $\Delta E_a = 2E_R/(n - \mu)^3$ and the corresponding difference in frequency is $\Delta\nu = \Delta E_a/h$. Since $\Delta\nu$ decreases like $(n - \mu)^{-3}$, transitions to individual states with n very large will not be resolved in any experiment with finite resolving power. In this region the quantity which can be measured is the mean cross-section defined by

$$\langle\sigma_\nu\rangle = \frac{1}{\Delta\nu} \int_{\Delta\nu} \sigma_\nu \, d\nu \qquad (18)$$

and for this we obtain

$$\langle\sigma_\nu\rangle = 4\pi^2 \alpha a_o^2 \, df/dE \qquad (19)$$

where $\alpha = e^2/(\hbar c)$ is the fine-structure constant ($\alpha \simeq 1/137$), $a_o = \hbar^2/(me^2)$ is the Bohr radius and

$$\frac{df}{dE} = \frac{(n - \mu)^3}{2} f \qquad (20)$$

is known as the differential oscillator strength per Rydberg. It can be shown that the mean cross-section (19) remains finite for $n \to \infty$ and joins smoothly to the cross-section for photo-ionisation (a detailed discussion is given by Dubau and Seaton, 1984). The expression for the photo-ionisation cross-section may be written

$$\sigma_\nu = 4\pi^2 \alpha a_o^2 \, df/dE \qquad (21)$$

where now

$$\frac{df}{dE} = \frac{(E_c - E_b)}{E_R} |D|^2/(3 g_b) \qquad (22)$$

and

$$D = (\Psi_c || \vec{R} || \Psi_a) \tag{23}$$

with the continuum-state function being normalised to

$$(\Psi_c(E) | \Psi_c(E')) = \delta(E - E') \tag{24}$$

with E and E' in Rydbergs.

Figure 5, from Pradhan and Saraph, shows df/dE for the $^3S^o$ and $^3D^o$ final states of oxygen in the regions of just below and just above the $^4S^o$ threshold. The series $(^4S^o)nd\ ^3D^o$ is seen to be perturbed by $(^2D^o)3s\ ^3D^o$, the interactions producing this perturbation being the same as those which give resonances in the continuum. Table 4 compares f-values of Pradhan and Saraph for the series $2p^4\ ^3P \rightarrow {}^3S^o$, $^3P^o$ and $^3D^o$ with values measured by Doering et al. (1985). The good agreement between the measured and calculated values suggests that the photo-ionisation cross-section should be correct to within about 10 per cent.

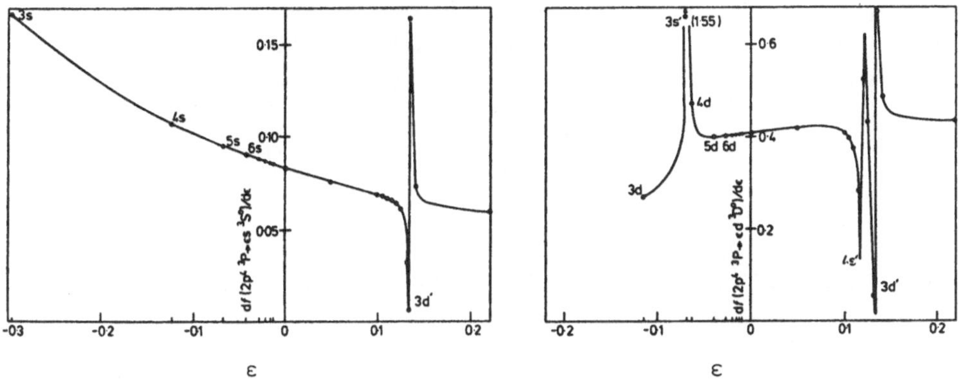

Figure 5. Differential oscillator strengths $df/d\varepsilon$ for the transitions in atomic oxygen $2p^4\ ^3P \rightarrow {}^3S^o$ (left-hand figure) and $2p^4\ ^3P \rightarrow {}^3D^o$ (right-hand figure). The energy ε is in Rydbergs and is relative to $O^+\ 2p^3\ ^4S^o$. From Pradhan and Saraph (1977).

Cross-sections for photo-ionisation contain series of resonances converging to each new threshold, which are similar to series of transitions to bound states converging to ionisation thresholds. The theory for averaging over resonances, to obtain mean cross-sections $\langle\sigma_\nu\rangle$ in the region below a new threshold, was first given by Gailitis (1963). For photo-absorption the mean cross-section just below a new threshold joins smoothly to the cross-section just above. The same is generally true for photo-ionisation but there are some exceptions. Figure 6(a), from Pradhan (1978), shows the photo-ionisation cross-section for atomic oxygen in the region below

Table 4. Oscillator strengths for transitions from the $2p^4\ ^3P$ ground state of atomic oxygen. Experimental values from Doering _et al._ (1985), calculated values from Pradhan and Saraph (1977).

Final state	f	
	Experiment	Calculated
$2p^3(^4S^o)3s\ ^3S^o$	0.048 ± 0.002	0.054
4s	0.010 ± 0.003	0.009
$2p^3(^4S^o)3s\ ^3D^o$	0.019 ± 0.002	0.020
$(^2D^o)3s$	0.061 ± 0.009	0.056
$(^4S^o)4d$	0.016 ± 0.007	0.015
$2p^3(^2P^o)3s\ ^3P^o$	0.086 ± 0.010	0.079
$2s2p^5$	0.070 ± 0.007	0.070

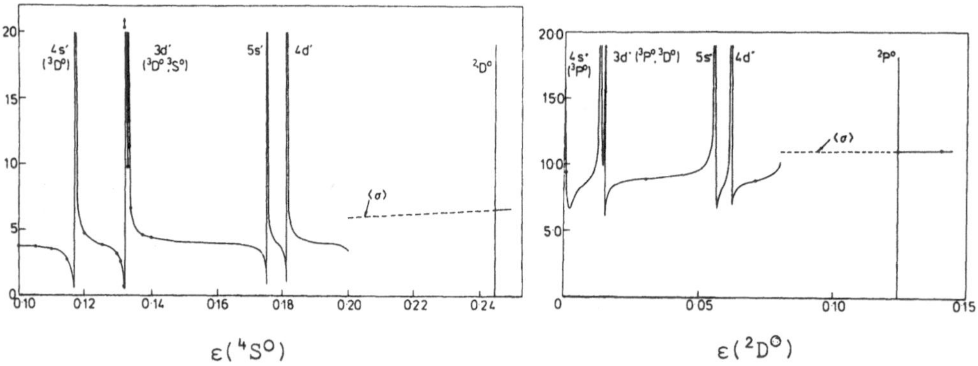

Figure 6. Atomic oxygen photo-ionisation cross-sections σ in Mb against energies ε in Rydbergs relative to thresholds indicated. From Pradhan (1978). Figure 6(a) shows $\sigma = \sigma(^4S^o)$ in the region below the $^2D^o$ threshold and Figure 6(b) shows $\sigma = \sigma(^4S^o) + \sigma(^2D^o)$ between the $^2D^o$ and $^2P^o$ thresholds. Cross-sections $\langle\sigma\rangle$ averaged over resonances are shown with dashed lines.

the $^2D^o$ threshold. The lower resonances are here resolved and the averaged cross-section, $\langle\sigma\rangle$, is shown by a dashed line. The results on this figure are for the sum of the contributions from $SL\pi = {}^3S^o$ and $^3D^o$. Figure 6(b) shows the cross-section in the region between the $^2D^o$ and $^2P^o$ thresholds, the sum of contributions from $SL\pi = {}^3S^o$, $^3P^o$ and $^3D^o$. There is seen to be a resonance, $(^2P^o)3s\ ^3P^o$, which is interrupted by the $^2D^o$ threshold. Figure 6(a) and (b) provides an example of a case for which the averaged photo-ionisation cross-section below a threshold does not join smoothly to that above. There is a smooth join in the contributions from $SL\pi = {}^3S^o$ and ${}^3D^o$ but in LS coupling there is no smooth join for ${}^3P^o$ because there are no continuum ${}^3P^o$ states below the ${}^2D^o$ threshold. There will be a smooth join in the photo-absorption cross-section if one includes the contributions from the $({}^2D^o)nd$ ${}^3P^o$ series. When allowance is made for departures from LS coupling, the ${}^3P^o$ states below the ${}^2D^o$ limit can auto-ionise and it is known from experiments that some of these states do auto-ionise. There is at present no information concerning the auto-ionisation widths for these states and the relative importance of auto-ionisation and radiative decays. One cannot therefore be certain about the extent to which these states contribute to photo-ionisation.

A detailed knowledge of the photo-ionisation process for atomic oxygen is required in ionospheric physics. The total rate of electron production depends only on the total cross-section,

$$\sigma = \sum_{\gamma} \sigma(\gamma) \tag{25}$$

where γ specifies a state of the O^+ ion produced, but the individual cross-sections $\sigma(\gamma)$ are required in order to calculate the energy distribution of the photo-electrons and the intensities in spectral lines which can be emitted following production of O^+ in excited states. Information concerning the angular distribution of the photo-electrons is also required. Since they are produced by uni-directional radiation from the sun, the photo-electrons have angular distributions which are anisotropic. In the F region electrons can travel for large distances, spiralling along the lines-of-force of the earth's magnetic field, and the trajectory of each electron depends on the angle between its initial direction and the magnetic-field direction.

At this point we require a more detailed description of the photo-ionisation process. Let $\Psi_{i'}$ be a final-state wave-function which has outgoing waves only in the channel i', which is referred to as the exit channel. This function has an expansion of the type (12) containing orbital functions $\theta_{ii'}$ with radial functions $F_{ii'}(r)$. The required normalisation and boundary conditions are satisfied on taking

$$F_{ii'}(r) \underset{r\to\infty}{\sim} \phi_i^+(r)\delta_{ii'} - \phi_i^-(r)S_{ii'}^* \tag{26}$$

where $\underset{\sim}{S}$ is the scattering matrix and

$$\phi_i^{\pm}(r) = (4\pi k_i)^{-1/2} \exp(\pm\, i\, \zeta_i) \tag{27}$$

where

$$\zeta_i = k_i r - \frac{1}{2} \ell_i \pi + (1/k_i) \ell n\, (2k_i r) + \eta_i \qquad (28)$$

and $\eta_i = \arg \Gamma(\ell_i + 1 - i/k_i)$. For an initial state b and exit channel $\gamma \ell S L \pi$ we define the reduced matrix-element

$$D(\gamma \ell S L \pi, b) = (\gamma \ell S L \pi || \vec{R} || b). \qquad (29)$$

The cross-section to produce ions in level γ is

$$\sigma(\gamma) = \frac{4\pi^2 \alpha a_0^2}{3 g_b} \frac{(E - E_b)}{E_R} \sum_{SL\pi} \sum_{\ell} \left| D(\gamma \ell S L \pi, b) \right|^2 . \qquad (30)$$

Let $I(\hat{k})d\omega$ be the cross-section for production of electrons in the direction $\hat{k}$ and within the solid angle $d\omega$. Using symmetry arguments it may be shown (Yang, 1948) that, for polarised radiation,

$$I = \frac{\sigma}{4\pi} \{1 + \beta P_2(\cos\theta)\} \qquad (31)$$

where θ is the angle between $\hat{k}$ and the electric vector of the radiation field; and that for uni-directional unpolarised radiation

$$I = \frac{\sigma}{4\pi} \{1 - \frac{1}{2} \beta P_2(\cos\theta)\} \qquad (32)$$

where now θ is the angle between $\hat{k}$ and the direction of radiation propagation. The parameter β depends on the detailed physics of the ionisation process and is discussed by, among others Lipsky (1967), Cooper and Zare (1969), Dill et al. (1974), Burke and Taylor (1975), E.R. Smith (1976) and Taylor (1977). In deriving the expression for β, assuming LS coupling, it is not necessary to specify S (which does not change) or π (which does). For radiation with electric vector in the Oz direction the operator $\vec{R}$ is replaced by its z-component, Z. Let the initial state b have orbital angular momentum quantum numbers $L_b M$ and let the final state be $L_i \ell L M$ where L_i is the orbital angular momentum of the ion produced. The matrix element

$$D_Z(L_i \ell L M, L_b M) = (L_i \ell L M | Z | L_b M) \qquad (33)$$

is obtained using the Wigner-Eckart theorem,

$$D_Z(L_i \ell L M, L_b M) = \frac{D(L_i \ell L, L_b)}{(2L+1)^{1/2}} C^{L_b 1 L}_{M 0 M} \qquad (34)$$

where $C^{abc}_{\alpha\beta\gamma}$ is a vector-coupling coefficient, and the matrix element $D_Z(L_i M_i \ell m, L_b M)$ is obtained from $D_Z(L_i \ell L M, L_b M)$ using vector-coupling formulae. For an initial state $L_b M$ the wave-function for the final state is proportional to

$$\sum_{L_i M_i} \sum_{\ell m} \psi(L_i M_i \ell m)\, D_Z(L_i M_i \ell m, L_b M). \qquad (35)$$

In order to obtain an expression for $I(\hat{k})$ one must collect together all contributions to the amplitude of the outgoing wave, $(1/r)\exp\{i(kr - (1/k)\ell n(2kr))\}$. The final expression for β is obtained after some Racah algebra (see, for example, Smith, 1976). This expression is not always quoted, or not always quoted correctly. It is

$$\beta \times \sum_{L\ell} |D(\ell L)|^2 = (-1)^{L_i+L_b} (30)^{1/2} \times \sum_{LL'} \begin{Bmatrix} 1 & 2 & 1 \\ L' & L_b & L \end{Bmatrix} \sum_{\ell\ell'} \begin{pmatrix} \ell & 2 & \ell' \\ 0 & 0 & 0 \end{pmatrix} \begin{Bmatrix} L' & 2 & L \\ \ell & L_i & \ell' \end{Bmatrix} E(\ell L)\, E^*(\ell' L') \tag{36}$$

where $D(\ell L) = D(L_i \ell L, L_b)$ and

$$E(\ell L) = \exp(i\eta) i^{-\ell} [(2\ell+1)(2L+1)]^{1/2} D(\ell L). \tag{37}$$

I return to a discussion of results for atomic oxygen photo-ionisation. Most, but not all, of the theory papers give the separate cross-sections $\sigma(^4S^o)$, $\sigma(^2D^o)$ and $\sigma(^2P^o)$. Branching ratios $\sigma(^2D^o)/\sigma(^4S^o)$ and $\sigma(^2P^o)/\sigma(^4S^o)$ were measured by Samson and Petrovsky (1974) using He I λ584 radiation and techniques of photo-electron spectroscopy. They compared their measured ratios with all of the calculated values available at the time, and found particularly good agreement between the measured ratios (of 1.57 ± 0.14 and 0.82 ± 0.07) and the results obtained by Henry (1967), (1.57 and 0.85), from CC calculations and the dipole-length formula. Further measurements of branching ratios have been made by Hussein et al.(1985) in the range 725 to 580 Å using a synchrotron light-source. They also obtained absolute cross-section on normalising to the data of Samson and Pareek.

Measurements of β have not, to my knowledge, been made for atomic oxygen although such measurements have been made for other atoms such as the rare gases. Calculations of β for oxygen have been made by Starace et al. (1974) and E.R. Smith (1976). The former use wave-functions calculated in both Hartree-Fock and Herman-Skillman potentials,and find that the differences between the values for these two cases are not large but are significant at the lower energies. Smith uses a CC theory similar to that of Henry (1967, 1968) and in most of his calculations only open channels are included, as in the 1967 work of Henry. Closed channels are included by Smith in a calculation of β in the region of the $(^2D^o)4s\ ^3D^o$ resonance. There are comparatively large differences between the values of β obtained by Smith using the length and velocity formulae, suggesting that β is rather sensitive to the accuracy of the functions used.

Results for the photo-ionisation of atomic oxygen can be summarised as follows.

(i) The cross-section is now known to an accuracy of 10%

(ii) Experimental work provides accurate values for the absolute value of the cross-section and for positions of resonances. There are no experimental results for the widths of the resonances.

(iii) Good absolute values for the calculated cross-section require the use of an accurate function Ψ_b for the initial state.

(iv) Good results for resonance structures require accurate functions Ψ_c for the final state.

References

Bates, D.R., 1939, Mon. Not. R. astr. Soc., 100, 25.
Bates, D.R., 1946a, Mon. Not. R. astr. Soc., 106, 423.
Bates, D.R., 1946b, Mon. Not. R. astr. Soc., 106, 432.
Bates, D.R., 1978, Phys. Rep., 35, 306.
Bates, D.R., Buckingham, R.A., Massey, H.S.W. and Unwin, J.J., 1939, Proc. R. Soc., A170, 322.
Bates, D.R., and Massey, H.S.W., 1940, Astrophys. J., 91, 202.
Bates, D.R., and Massey, H.S.W., 1941, Proc. R. Soc., A177, 281.
Bates, D.R., and Seaton, M.J., 1949, Mon. Not. R. Astr. Soc., 106, 699.
Burke, P.G., Hibbert, A., and Robb, W.D., 1971, J. Phys. B; At. Molec. Phys., 11, 143.
Burke, P.G., and Taylor, K.T., 1975, J. Phys. B; At. Molec. Phys., 8, 2620.
Cairns, R.B., and Sampson, J.A.R., 1965, Phys. Rev., A139, 1403.
Comes, F.J., Speier, F., and Elzer, A., 1968, Z. Naturf., 23a, 125.
Cooper, J., and Zare, R.N., 1969, Lecture in Theoretical Physics (ed. S. Geltman, K.T. Mahanthappa and W.E. Brittin), Vol.11, p317.
Dalgarno, A., Henry, R.J.W., and Stewart, A.I., 1964, Plan. Sp. Sci., 12, 235.
Dalgarno, A., and Lewis, J.T., 1956, Proc. Phys. Soc., A69, 285.
Dalgarno, A., and Parkinson, D., 1960, J. Atmos. Terr. Phys., 18, 335.
Dehmer, P.M., Berkowitz, J., and Chupka, W.A., 1973, J. Chem. Phys., 59, 5777.
Dill, D., Manson, S.T., and Starace, A.F., 1974, Phys. Rev., A11, 1596.
Doering, J.P., Gulcicek, E.E., and Vaughan, S.O., 1985, J. Geophys. Res., 90, 5279.
Dubau, J., and Seaton, M.J., 1984, J. Phys. B; At. Molec. Phys., 17, 381.
Gailitis, M., 1963, Sov. Phys. - JETP, 17, 1328.
Ganas, P.S., 1973, Phys. Rev., A7, 928.
Hartree, D.R., and Black, M.M., 1933, Proc. Roy. Soc., A139, 311.
Hartree, D.R., Hartree, W., and Swirles, B., 1939, Phil. Trans. Roy. Soc., A238, 229.
Henry, R.J.W., 1967, Plan. Sp. Sci., 15, 1747.
Henry, R.J.W., 1968, Plan. Sp. Sci., 16, 1503.
Herman, F., and Skillman, S., 1963, Atomic Structure Calculations, Prentice-Hall.
Huffman, R.E., Larrabee, J.C., Tanaka, Y., 1966, J. Chem. Phys., 46, 2213.

Hussein, M.I.A., Holland, D.M.P., Codling, K., Woodruff, P.R., and Ishiguro, E., 1985, J. Phys. B; At. Molec. Phys., 18, 2827.
Kähler, H., 1971, J. Quant. Spect. Rad. Transf., 11, 1521.
Kohl, J.L., Lafyatis, G.P., Palenius, H.P., and Parkinson, W.H., 1978, Phys. Rev., A18, 571.
Lipsky, L., 1967, Abstract of 5th Int. Conf. Phys. Electronic and atomic collisions (I.P. Flacks and E.S. Solovyov), p.617. Nauka, Leningrad.
McGuire, E.J., 1968, Phys. Rev., A175, 20.
Manson, S.T., Msezane, A., and Starace, A.F., 1979, Phys. Rev., A20, 1005.
Moore, C.E., 1976, Selected Tables of Atomic Spectra, NSRDS-NBS 3 Section 7, Nat. Bureau Stds., Washington.
Pradhan, A.K., 1976, Ph.D. Thesis, University of London.
Pradhan, A.K., 1978, J. Phys. B; At. Molec. Phys., 11, L729.
Pradhan, A.K., and Saraph, H.E., 1977, J. Phys. B; At. Molec. Phys., 10, 3365.
Rudd, M.E., and Smith, K., 1968, Phys. Rev. A169, 79.
Samson, J.A.R., and Pareek, P.N., 1985, Phys. Rev., A31, 1470.
Samson, J.A.R., and Petrovsky, V.E., 1974, Phys. Rev., A9, 2449.
Seaton, M.J., 1953a, Phil. Trans. R. Soc., A245, 469.
Seaton, M.J., 1953b, Proc. Roy. Soc., A218, 400.
Seaton, M.J., 1974, J. Phys. B; At. Molec. Phys., 7, 1817.
Seaton, M.J., 1982, Q. Jour. R. Astr. Soc., 23, 2.
Smith, E.R., 1976, Phys. Rev., A13, 1058.
Starace, A.F., Manson, S.T., and Kennedy, D.J., 1974, Phys. Rev., A9, 2453.
Taylor, K.T., 1976, Ph.D. Thesis, The Queen's University of Belfast.
Taylor, K.T., 1977, J. Phys. B; At. Molec. Phys., 10, L699.
Taylor, K.T., and Burke, P.G., 1978, J. Phys. B; At. Molec. Phys., 9, L353.
Thomas, G.M., and Helliwell, T.M., 1970, J. Quant. Spect. Rad. Transf., 10, 423.
Yamanouchi, T., and Kotani, M., 1940, Proc. Math. Phys. Soc., Japan, 22, 60.
Yang, C.N., 1948, Phys. Rev., 74, 764.

THE FORMATION OF COMPLEX INTERSTELLAR MOLECULES

A. Dalgarno

Harvard-Smithsonian Center for Astrophysics
Cambridge
Massachusetts

1. INTRODUCTION

A rich array of complex molecules has been detected in interstellar space. Table 1 is a list arranged in order by the number of atoms contained in the molecule. For the complex organic species the abundances relative to hydrogen are of the order 10^{-8}-10^{10} or less. The abundances vary from one interstellar cloud to another. In TMC-1 there are observed high abundances of cyanopolyynes and complex hydrocarbons, whereas in SGR B2 oxygen-bearing organic molecules appear to be relatively enhanced. In the Orion molecular cloud, large variations occur which are associated with high velocity outflows from embedded young stars. Various attempts have been made to construct theoretical models of the formation and destruction of interstellar molecules. We review them here with particular reference to molecular species containing several carbon atoms.

2. THEORETICAL MODELS

In a dense interstellar cloud, the gas is composed of hydrogen molecules and helium atoms with a small admixture of heavier elements, principally oxygen, carbon, nitrogen, sulphur and silicon. The gas contains solid particles of uncertain composition which in a dense cloud absorb out the external interstellar radiation field. The particle densities range from 10^3 cm $^{-3}$ to 10^6 cm $^{-3}$ and the temperatures from 10K to 50K, though significant warming may occur in local regions. The clouds are subjected to galactic cosmic rays with an ionization rate usually assumed to be about $10^{-17}s^{-1}$.

The theoretical models of the chemistry of such clouds all have the same general structure though they differ in important ways in the role attributed to grain surface chemistry. Reactions between neutral particles are often inhibited at low temperatures and the gas phase chemistry of cold clouds is an ion-molecule chemistry mediated by a few neutral particle reactions with zero or low activation energies (cf. Smith 1987).

The carbon gas phase chemistry is initiated by the radiative association of C^+ and H_2,

$$C^+ + H_2 \rightarrow CH_2^+ + h\nu \ , \qquad (1)$$

with a rate coefficient taken to be about $5.10^{-16}cm^3s^{-1}$ because that is the value obtained empirically from models which reproduce the measured abundances of CH in diffuse clouds. A value near $5.10^{-16}cm^3s^{-1}$ has some theoretical support, but remains a crucial uncertainty in interstellar chemistry.

TABLE 1

Interstellar Molecules

H_2	Hydrogen	CH	Methylidyne
CH^+	Methylidyne Ion	OH	Hydroxyl
C_2	Carbon	CN	Cyanogen
CO	Carbon Monoxide	NO	Nitric Oxide
CS	Carbon Monosulphide	SiO	Silicon Monoxide
SO	Sulphur Monoxide	NS	Nitrogen Sulfide
SiS	Silicon Sulphide	HCl	Hydrogen Chloride
H_2O	Water	C_2H	Ethynyl
HCN	Hydrogen Cyanide	HNC	Hydrogen Isocyanide
HCO	Formyl	HCO^+	Formyl ion
N_2H^+	Protonated Nitrogen	H_2S	Hydrogen Sulphide
HNO	Nitroxyl	OCS	Carbonyl Sulphide
SO_2	Sulphur Dioxide	HCS^+	Thioformyl ion
NH_3	Ammonia	H_2CO	Formaldehyde
HNCO	Isocyanic Acid	H_2CS	Thioformaldehyde
C_3N	Cyanoethynyl	HNCS	Isothiocyanic Acid
C_3H	Propynlidyne	$HOCO^+$	Protonated Carbon Dioxide
C_3O	Tricarbon Monoxide	$HCNH^+$	Protonated Hydrogen Cyanide
CH_2CO	Ketene	C_3H_2	Cyclopropenylidene
HCOOH	Formic Acid	HCOOH	Formic Acid
C_4H	Butadinyl	HC_3N	Cyanoacetylene
CH_2NH	Methanimine	NH_2CH	Cyanamide
C_5H	Pentynylidyne	CH_3CH	Methyl Cyanide
CH_3OH	Methyl Alcohol	CH_3SH	Methyl Mercaptan
C_6H	Hexatriynl	CH_3C_2H	Methyl Acetylene
NH_2CHO	Formamide	CH_2CHCN	Vinyl Cyanide
CH_3NH_2	Methylamine	$HCOOCH_3$	Methyl Formate
CH_3CHO	Acetaldehyde	HC_5N	Cyanodiacetylene
CH_3C_3N	Methyl Cyanoacetylene	CH_3C_4H	Methyl Diacetylene
CH_3CH_2OH	Ethyl Alcohol	$(CH_3)_2O$	Dimethyl Ether
HC_7N	Cyanohexatriyne	CH_3CH_2CN	Ethyl Cyanide
HC_9N	Cyano-octatetra-yne	CH_3C_4H	Methyl Diacetylene
$HC_{11}N$	Cyano-decapentra-yne		

The reaction is followed by the rapid abstraction reaction

$$CH_2^+ + H_2 \rightarrow CH_3^+ + H \qquad (2)$$

In dense clouds where the electron density is low, CH_3^+ is removed only slowly by dissociative recombination and it also participates in radiative association reactions with other constituents Y,

$$CH_3^+ + Y \rightarrow CH_3^+ . Y + h\nu \qquad (3)$$

to build more complex ions. The complex ions may undergo dissociative recombination, which is usually assumed to proceed on one or both of the branches

$$CH_3^+ \cdot Y + e \rightarrow CH_2Y + H$$
$$\rightarrow CHY + H_2 \ , \quad (4)$$

producing the neutral molecules CH_2Y and CHY. Thus, methanol may be produced by the sequence

$$CH_3^+ + H_2O \rightarrow CH_3^+ \cdot H_2O + h\nu \quad (5)$$

$$CH_3^+ \cdot H_2O + h\nu \rightarrow CH_3\,OH + H \quad (6)$$

and dimethyl ether by the sequence

$$CH_3^+ \cdot CH_3\,OH \rightarrow CH_3^+ \cdot CH_3\,OH + h\nu \quad (7)$$

$$CH_3^+ \cdot CH_3\,OH + e \rightarrow CH_3\,OCH_3 + H \quad (8)$$

The CH_3^+ ions also participate in a radiative association reaction with the major constituent H_2 to form CH_5^+,

$$CH_3^+ + H_2 \rightarrow CH_5^+ + h\nu \ . \quad (9)$$

By reacting with CO, CH_5^+ is a source of methane

$$CH_5^+ + CO \rightarrow CH_4 + HCO^+ \ . \quad (10)$$

Condensation reactions such as

$$CH_3^+ + CH_4 \rightarrow C_2H_5^+ + H_2 \quad (11)$$

may then occur. An important fraction of the CH_3^+ ions is removed by

$$CH_3^+ + O \rightarrow HCO^+ + H_2 \quad (12)$$

which leads to CO by dissociative recombination

$$HCO^+ + e \rightarrow H + CO \quad (13)$$

and by reactions with H_2O,

$$HCO^+ + H_2O \rightarrow H_3O^+ + CO \ . \quad (14)$$

There are other sequences that lead to CO and the models all show that in the steady-state in clouds with a gas phase carbon abundance [C] less than the oxygen abundance [O], most of the carbon is locked up in CO with an abundance ratio $n(CO)/n_H$ of the order 10^{-4}. None of the other carbon-bearing species has a steady-state abundance ratio exceeding 10^{-7}. We consider later the modifications to interstellar chemistry that arise if the interstellar gas contains a component of polycyclic aromatic hydrocarbons with an abundance ratio near 10^{-7}.

Some of the CO is converted into other molecules by the extraction of C^+ through the raction with He^+,

$$He^+ + CO \rightarrow He + C^+ + O \ . \quad (15)$$

Because He^+ reacts slowly with H_2 at low temperatures, reaction (15) is the main He^+ removal mechanism. Following reaction (15), molecular species containing several carbon atoms may be built by carbon insertion reactions involving C^+, such as

$$C^+ + CH_4 \rightarrow C_2H_3^+ + H \quad (16)$$
$$C^+ + C_2H_2 \rightarrow C_3H^+ + H \ . \quad (17)$$

Parallel to the carbon chemistry there is an oxygen chemistry driven by the proton transfer

$$H_3^+ + O \rightarrow OH^+ + H_2 \quad (18)$$

the H_3^+ arising from cosmic ray ionization of H_2 followed by

$$H_2^+ + H_2 \rightarrow H_3^+ + H \ . \quad (19)$$

Reaction (18) initiates the abstraction sequence

$$OH^+ + H_2 \rightarrow H_2O^+ + H \quad (20)$$

$$H_2O^+ + H_2 \rightarrow H_3O^+ + H \ . \quad (21)$$

The molecular ion H_3O^+ is very stable. It is removed by dissociative recombination

$$H_3O^+ + e \rightarrow H_2O + H \quad (22)$$

$$\rightarrow OH + H_2 \ , \quad (23)$$

by proton transfer in reactions with some molecular species, such as

$$H_3O^+ + C_3H \rightarrow C_3H_2^+ + H_2O \ , \quad (24)$$

and by radiative association processes. The radiative association

$$H_3O^+ + C_2H_4 \rightarrow CH_3CH_2OH_2^+ + h\nu \ , \quad (25)$$

followed by

$$CH_3CH_2OH_2^+ + e \rightarrow CH_3CH_2OH \quad (26)$$

may be the most rapid route for the formation of ethyl alcohol.

The carbon and oxygen chemistries are linked by reactions (10) and (11) and by

$$C^+ + H_2O \rightarrow HCO^+ + H \quad (27)$$

and

$$C^+ + O_2 \rightarrow CO^+ + O \ . \quad (28)$$

A similar chemistry can be constructed for nitrogen-bearing compounds. Thus

$$CH_3^+ + HCN \rightarrow CH_3^+ \ . \ HCN + h\nu \quad (29)$$

$$CH_3^+ \ . \ HCN + e \rightarrow CH_3CN + H \quad (30)$$

produces the interstellar molecule, methyl cyanide. The cyanopolynes HC_nN can be produced by reactions of N and HCN with hydrocarbon ions. The preferential route to HC_3N appears to be

$$C_3H_3^+ + N \rightarrow H_2C_3N^+ + H \quad (31)$$

$$H_2C_3N^+ + e \rightarrow HC_3N + H \quad (32)$$

(Knight et al. 1986).

With this structure of ion-molecule reactions, dissociative recombination and radiative association, schemes can be devised that lead to the gas phase production of the observed interstellar molecules (Huntress and Mitchell 1979, Mitchell and Huntress 1979, Mitchell, Huntress and Prasad 1979, Prasad and Huntress 1980ab, Herbst 1983, Leung, Herbst and Huebner 1984, Millar and Freeman 1984, Millar and Nejad 1985, Herbst and Leung 1986a,b, Millar, Leung and Herbst 1987, Herbst 1987). However serious difficulties confront attempts to reproduce the measured abundances of complex interstellar molecules so that useful conclusions can be drawn from a comparison of theory and observationa. There are major uncertainties, of both quantitative and a qualitative nature, in the description of the chemical network.

3. MODEL RESULTS

The uncertainties in the reaction rate coefficients permit large differences in the calculated abundances of complex species. In Table 2, a comparison is given of some of the steady-state results of Millar and Nejad

TABLE 2

Calculated abundance ratios $n(x)/n(H_2)$

X	M-N	H-L	H-L'	SDL	SDL'
OH	7.2(-8)	6.5(-7)	2.5(-7)	5.6(-8)	5.6(-8)
H_2O	6.8(-8)	3.6(-6)	1.3(-6)	1.4(-6)	1.4(-6)
H_3O^+	2.3(-10)	1.5(-9)	2.0(-9)	1.3(-9)	1.3(-9)
C^+	3.7(-9)	1.1(-9)	1.3(-9)	6.2(-10)	6.2(-10)
H_3^+	5.4(-10)	3.4(-9)	1.8(-9)	3.3(-9)	3.4(-9)
CH_4	1.6(-7)	6.8(-8)	1.1(-7)	5.0(-8)	4.9(-8)
C_2H	6.2(-9)	2.4(-10)	1.7(-10)	1.1(-9)	1.7(-9)
C_2H_2	2.6(-8)	3.8(-9)	9.9(-9)	8.0(-10)	2.1(-9)
C_3H	1.2(-7)	8.5(-10)	6.5(-11)	1.4(-10)	2.9(-10)
C_3H_2	7.7(-8)	4.4(-10)	8.5(-10)	7.2(-11)	1.5(-10)

M-N Millar and Nejad (1985), H-L Herbst and Leung (1986a), H-L' Herbst and Leung (1986b), SDL Sternberg, Dalgarno and Lepp (1987) with f = 0.5, SDL' Sternberg et al. with reactions (39-45).

(1985) and Herbst and Leung (1986a), who appear to apply similar chemical considerations to similar physical environments. There are differences in detail, particularly in the treatment of reactions involving atomic oxygen and in branching ratios of dissociative recombination, which are sufficient to bring about an abundance ratio for C_3H of $1.2 \cdot 10^{-7}$ in one model and $8.5 \cdot 10^{-10}$ in the other.

The differences have been analyzed by Millar, Leung and Herbst (1987). In the model of Herbst and Leung (1986a), C_3H is most rapidly removed by

$$C_3H + HCO^+ \rightarrow C_3H_2^+ + CO \ . \qquad (33)$$

The $C_3H_2^+$ ion does not react with H_2 and it is removed by

$$C_3H_2^+ + e \rightarrow C_3H + H \qquad (34)$$

$$\rightarrow C_3 + H_2 \ . \qquad (35)$$

Triatomic carbon is then quickly destroyed by atomic oxygen in the reaction

$$C_3 + O \rightarrow C_2 + CO \qquad (36)$$

so that there is a net loss of C_3H. In the model of Millar and Nejad (1986), reaction (35) is treated as negligible and reaction (33) is followed by (34) with no loss of C_3H.

The identity of the end products of the dissociative recombination of $C_3H_2^+$ is in fact unknown. There are two conflicting theoretical views. According to Herbst (1978) and Green and Herbst (1979), the channels are (34) and (35) in some uncertain proportion. According to Bates (1950, 1986, 1987) the channels are (34) and

$$C_3H_2^+ + e \rightarrow C_2 + CH_2 \qquad (37)$$

$$\rightarrow C + C_2H_2 \qquad (38)$$

(Millar, Leung and Herbst 1987). The breaking of the carbon bond in dissociative recombination is a formidable obstacle to the formation of complex hydrocarbons by the association of simpler systems.

Although it had long been recognized that the rate coefficients for the reactions of ions with heteronuclear molecules at low temperatures often exceed the Langevin values thought to be appropriate to the reactions of ions with homonuclear molecules (cf. Dalgarno 1970, Su and Bowers 1975, 1979, Takayanagi 1978), the enhanced rate coefficients were not incorporated into the early models of interstellar molecular clouds, perhaps because of other larger uncertainties that attend the models. However, it was demonstrated by Millar et al. (1985) that the effects could be substantial. Theoretical studies (cf. Bates 1987) and laboratory measurements (Clary, Smith and Adams 1985, Marquette et al. 1985) have provided more reliable values of the rate coefficients. They have been included in a study by Herbst and Leung (1986b), whose results are reproduced in column H-L' of Table 2.

The effects on the abundances are illustrated by a comparison of the columns H-L and H-L'. The abundances of the polar molecules are reduced and those of the non-polar molecules enhanced, but the changes are less than an order of magnitude.

Similar calculations have been carried out by Sternberg, Dalgarno and Lepp (1987). Their results are reproduced in column SDL of Table 2. Agreement with the abundances calculated by Herbst and Leung (1986b) is satisfactory except for C_2H and C_2H_2 where the discrepancies are an order of magnitude in opposite directions.

The discrepancies illustrate the sensitivity of the predictions to the adopted chemistry. Herbst and Leung (1986b) included the reactions

$$C_2H_2^+ + H_2 \rightarrow C_2H_4^+ + h\nu \tag{39}$$

$$C_2H_4^+ + e \rightarrow C_2H_3 + H \tag{40}$$

$$\rightarrow C_2H_2 + H_2 \tag{41}$$

$$HCO^+ + C_2H_3 \rightarrow C_2H_4^+ + CO \tag{42}$$

$$H_3^+ + C_2H_3 \rightarrow C_2H_4^+ + H_2 \tag{43}$$

$$C^+ + C_2H_3 \rightarrow C_3H_2^+ + H \tag{44}$$

$$\rightarrow C_3H^+ + H_2 \tag{45}$$

whereas the results labelled SDL were obtained excluding them. The column SDL' in Table 1 contains the results of Sternberg et al. (1987) when the reactions (39)-(45) are retained in the chemical scheme. The only significant difference from the abundances calculated by Herbst and Leung (1986b) occurs for ethynyl and stems from a different choice of the rate coefficient for the reaction

$$C_2H + O \rightarrow CO + CH. \tag{46}$$

The close agreement, otherwise, of the two calculations demonstrates that similar assumptions lead to similar results but provides no evidence on the correctness of the assumptions. One assumption, common to these models, is that the fraction of the dissociative recombinations of H_3O^+

$$H_3O^+ + e \rightarrow H_2O + H \tag{47}$$

$$\rightarrow OH + H_2 \tag{48}$$

that produces H_2O is 0.5 as is the fraction f that produces OH. Bates (1986) has argued that all recombinations of H_3O^+ produce H_2O. The changes in abundances that result (Sternberg et al. 1987) are shown in Table 3 for the simpler model in which reactions (39)-(45) were not included.

The abundance of OH is drastically reduced, that of H_2O much enhanced. Because C^+ is removed by reaction with H_2O, its abundance is decreased. The abundance of H_3O^+ is increased above that of H_3^+ and proton transfer from H_3O^+ drives the chemistry. However, the abundances of the hydrocarbons are little changed (Sternberg et al. 1987).

The OH abundance is still well below the measured values. Though some OH molecules can be supplied by reactions of oxygen with components of the nitrogen and sulphur chemistries if favorable assumptions are made about the

TABLE 3

Abundances for branching ratios f=0.5 and 0

	f=0.5*	f=0*	ω=0.5
OH	5.6(-8)	8.6(-10)	1.3(-8)
H_2O	1.4(-6)	9.1(-5)	2.9(-5)
C^+	6.2(-10)	1.1(-10)	2.5(-10)
H_3O^+	1.3(-9)	8.0(-9)	6.3(-9)
H_3^+	3.3(-9)	1.1(-9)	2.1(-9)
CH_4	5.0(-8)	5.3(-8)	1.6(-8)
C_2H_2	7.9(-10)	2.5(-9)	2.2(-10)
C_3H	1.4(-10)	1.2(-10)	1.2(-11)
C_3H_2	7.2(-11)	6.6(-11)	6.0(-12)

* These calculations ignore the internally generated photons.

products of dissociative recombination of nitrogen and sulphur-bearing ions and of chemical reactions with oxygen, a simpler explanation may lie with the ultraviolet photons created internally by cosmic ray ionizing events (Prasad and Tarafdar 1983). Lepp, Dalgarno and Sternberg (1987) have put forward a quantitative analysis which indicates indeed that photodissociation of H_2O is the major source of OH in dense clouds. Photodissociation and photoionization by the cosmic ray induced photons have been included in a model of interstellar chemistry by Sternberg et al. (1987). The rates of photodissociation and photoionization depend on the grain albedo ω in the ultraviolet. The results of Sternberg et al. (1987) for ω= 0.5 and f=0 are shown in Table 3.

The OH abundance is increased and it appears that with the addition of photodissociaton, the OH abundance is insensitive to the branching ratio for dissociative recombination. The hydrocarbons are diminished by an order of magnitude. Because the destructive effects tend to amplify as more carbon atoms are added, the cosmic ray photons may pose grave problems in forming a sufficient abundance of large molecules.

The problems can be readily ameliorated by the arbitrary hypothesis that [C] / [O] exceeds unity so that excess carbon atoms remain after CO has formed (Langer et al. 1984, Watt 1985, Herbst and Leung 1986a). The hypothesis provides a ready explanation of the large C/CO ratio that has been observed in some dark clouds (Keene et al. 1985). However clouds may not be in a steady-state and the equilibrium values presented in the tables may be inappropriate.

The time-dependent behaviour of simple species has been explored by Oppenheimer and Dalgarno (1975), Iglesias (1977), Henning (1981), Graedel, Langer and Frerking (1982) and Watt (1983, 1985) and of complex systems also by Prasad and Huntress (1980a,b), Leung et al. (1984), Millar and Nejad (1985), Herbst and Leung (1986a,b), Brown and Rice (1986), and Millar, Leung and Herbst (1987). In a cloud of uniform density $10^4 cm^{-3}$ in which the carbon exists initially as C^+, the conversion of carbon into carbon monoxide is not completed until a time of 10^6 Years has elapsed and carbon chain molecules achieve abundances at early times that are much higher than their steady-state values. The chemical evolution will be accelerated by the large rate coefficients appropriate to reactions of positive ions with heteronuclear molecules and the peak abundances may be enhanced (Herbst and Leung 1986b). Chemical evolution may also be accelerated by the increased cooling of the gas due to molecular formation and the accompanying increase in density (Oppenheimer and Dalgarno 1975) and generally by dynamical effects (Gerola and Glassgold 1978, Tarafdar et al. 1985).

The steady-state calculations reported in Tables 2 and 3 do not take into account the loss of gas phase material by accretion into grain surfaces nor the release of molecules from the grains into the gas. There is observational evidence that a large fraction of interstellar CO exists in solid form (Lacy et al. 1984, Whittet et al. 1985, Larson et al. 1985, Gebelle 1986). The time-scale for accretion in a typical dense cloud is about 10^5 years (Iglesias 1977, de Jong, Dalgarno and Boland 1980, Millar and Nejad 1985). There appear to be few, if any, dense clouds, devoid or nearly so, of emitting molecules and containing only H_2, H_3^+ and HeH^+ in molecular form, so that mechanisms must be operating which inject material from the grain mantles into the gas. The nature of the mechanisms and the chemical composition of the material may exert a large influence on the molecular abundances in the cloud.

Various mechanisms have been suggested including photodesorption (Boland and de Jong 1982), heavy cosmic rays (Léger, Jura and Omont 1985), grain-grain collisions (d'Hendecourt, Allamandola and Greenberg 1985) and shocks (Williams and Hartquist 1984).

A model has been developed by d'Hendecourt et al. (1985) in which there is a continuous exchange of material between the solid and gas phases. A different model has been explored by Charnley et al. (1987) in which material is injected back into the gas phase by discrete events. Neither of these studies addresses the question of the formation of complex species.

There may be in any case a missing element in all the models used in the calculation of the abundances. It has been suggested that polycyclic aromatic hydrocarbons, PAH, form a component of the interstellar gas with a relative abundance between 10^{-7} and 10^{-6} (Léger and Puget 1984, Allamandola, Tielens and Baker 1985, van der Zwe and Allamandola 1985, Léger and d'Hendecourt 1985, Crawford, Tielens and Allamandola 1985, d'Hendercourt et al. 1986), though questions have been raised as to their survival against attack by oxygen and hydrogen (Duley and Williams 1986).

If they do exist in high abundance, the negative charge in a dense cloud is carried not by electrons but by negative PAH^- ions (Omont 1986, Lepp and Dalgarno 1987). The chemistry is qualitatively changed and dissociative recombination is largely replaced by mutual neutralization

$$M^+ + PAH^- \rightarrow \text{neutral products} . \qquad (49)$$

Table 4 is adapted from calculations by Lepp and Dalgarno (1987). In it are listed abundance ratios calculated without PAH molecules and with PAH molecules at a total abundance ratio of 10^{-7} for the cloud model of Table 3. The effects are large. The electron density is made negligible as the negative charge becomes PAH^-. The abundance of neutral carbon is enhanced by two orders of magnitude through the reaction

$$C^+ + PAH^- \rightarrow C + PAH . \qquad (50)$$

The major reaction sources of the hydrocarbons are altered and the hydrocarbon abundances are increased. There is a qualitative change in the pattern of reactions, whose consequences have yet to be explored. Because the electron density is decreased by attachment to PAH, the abundance of H_3O^+, which in the absence of PAH is removed by dissociative recombination, is considerably larger. It may provide an upper limit to the fractional abundance of PAH molecules.

The disruption of PAH molecules by Auger processes triggered by cosmic rays and X-rays is a potentially significant additional source of molecular species leading to complex molecules (Buch, Lepp and Dalgarno 1987).

In conclusion, it seems we do not know yet with any certainty how the complex interstellar molecules are formed and we cannot use their existence as a reliable diagnostic probe of the astronomical events that created then and whose evolution they influence. But much progress has been made, brought about in large part by a deeper understanding of the mechanisms underlying the various kinds of chemical reactions, to which Sir David Bates continues to make notable contributions (cf. Bates 1987).

TABLE 4

Abundances $n(x)/n_H$ corresponding to an ionization rate of $10^{-17} s^{-1}$ and a fractional PAH abundance of zero and 10^{-7} *

$n(PAH)n_H$	0	10^{-7}
n_e	4.4(-8)	4.2(-9)
C^+	2.6(-10)	2.6(-10)
C	7.8(-10)	2.1(-7)
C_2H	1.1(-10)	1.8(-9)
C_2H_2	2.1(-10)	5.5(-9)
C_3H	1.1(-11)	1.8(-10)
C_3H_2	5.8(-12)	4.1(-10)

* From Lepp and Dalgarno (1987).

ACKNOWLEDGMENTS

I am indebted to Dr S. Lepp for instructive discussions. The research has been partly supported by the National Science Foundation under Grant AST-86-17975.

REFERENCES

Allamandola, L.J., Tielens, A.G.G.M. and Barker, J.R., 1985, Polycyclic aromatic hydrocarbons and the unidentified infrared emission bands: Auto exhaust along the Milky Way. Astrophys.J.Lett. 290, L25.

Bates, D.R., 1986, Products of dissociative recombination of polyatomic ions. Astrophys.J.Lett. 306, L45.

Bates, D.R., 1987, Interstellar clouds chemistry revisited in Recent Studies in Atomic and Molecular Processes (ed. A.E. Kingston) Plenum Press, London.

Bates, D.R., 1950, Dissociative recombination. Phys.Rev. 78, 492.

Boland, W. and de Jong, T., 1980, Carbon depletion in turbulent molecular cloud cores. Astrophys.J. 261, 110.

Brown, R.D. and Rice, E.H.N., 1986, Galactochemistry - 1. Influence of initial conditions on predicted abundances. Mon.Not.Roy.Astro.Soc. 223, 405.

Buch, V., Lepp, S. and Dalgarno, A., 1987, in preparation.

Charnley, S.B., Dyson, J.E., Hartquist, T.W. and Williams, D.A., 1987, Theoretical models of mass loaded flows: II. The chemistry in T-Tauri wind blown bubbles in dense molecular clouds. Mon.Not.Roy.Soc. in press.

Clary, D.C., Smith, D. and Adams, N.G., 1985, Temperature dependence of rate coefficients for reactions of ions with dipolar molecules. Chem.Phys. Lett. 119, 320.

Crawford, M.K., Tielens, A.G.G.M. and Allamandola, L.J., 1985, Ionized polycyclic aromatic hydrocarbons and the diffuse interstellar bands. Astrophys.J.Lett. 293, L45.

Dalgarno, A., 1970, Theory of ion-molecule collisions in ion-molecule reactions. E.W. McDaniel, V. Cermak, A. Dalgarno, E.E. Ferguson and L. Friedman Wiley, New York.

de Jong, T., Dalgarno, A. and Boland, W., 1980, Hydrostatic models of molecular clouds 1. Steady state models. Astron. Ap. 91, 68.

d'Hendecourt, L.B., Allamandola, L.J. and Greenberg, J.M., 1985, Time-dependend chemistry in dense interstellar clouds 1. Grain surface reactions, gas/grain interactions and infrared spectroscopy. Astron.Ap. 152, 130.

d'Hendecourt, L.R., Léger, A., Olofson, G. and Schmidt, W., 1986, The red rectangle: a possible case of visible luminescence from polycyclic aromatic hydrocarbons. Astron.Ap. 170, 91.

Geballe, T.R., 1986, Absorption by solid and gaseous CO towards obscured infrared objects. Astron.Ap. 162, 248.

Gerola, H. and Glassgold, A.E., 1978, Molecular evolution of contracting clouds: basic methods and initial results. Astrophys.J.Suppl. 37, 1.

Graedel, T.E., Langer, W.D. and Frerking, M.A., 1982, The kinetic chemistry of dense interstellar clouds. Astrophys.J.Suppl. 48, 321.

Green, S. and Herbst, E., 1979, Metastable isomers: a new class of interstellar molecules. Astrophys.J. 229, 121.

Henning, K., 1981, Molecular formation in interstellar clouds by gas phase reactions. Astron. Ap. Suppl. 44, 405.

Herbst, E., 1978, What are the products of polyatomic ion-electron dissociative recombination reactions? Astrophys. J. 222, 508.

Herbst, E., 1983, Ion-molecule synthesis of interstellar molecular hydrocarbons through C_4H: toward molecular complexity. Astrophys. J. Suppl. 53, 41.

Herbst, E., 1987, Can gas phase reactions produce complex oxygen-containing molecules in dense interstellar clouds? A revision of some important radiative association rate coefficients. Astrophys. J. in press.

Herbst, E. and Leung, C.M., 1986a, Synthesis of complex molecules in dense interstellar clouds in a gas-phase chemistry: model update and sensitivity analysis. Mon.Not.Roy.Astron.Soc. 222, 689.

Herbst, E. and Leung, C.M., 1986b, Effects of large rate coefficients for ion-polar neutral reactions on chemical models of dense interstellar clouds. Astrophys. J. 310, 378.

Huntress, W.T. and Mitchell, G.F., 1979, The synthesis of complex molecules in interstellar clouds. Astrophys. J. 208, 237.

Iglesias, E., 1977, The chemical evolution of molecular clouds. Astrophys. J. 218, 697.

Keene, J., Blake, G.A., Phillips, T.G., Huggins, P.J. and Buchman, C.A., 1985, The abundance of atomic carbon near the ionization fronts in M17 and S140. Astrophys. J. 299, 967.

Knight, J.S., Freeman, C.G., McEwan, M.J., Smith, S.C., Adams, N.G. and Smith, D., 1986, Production and loss of HC_3N in interstellar clouds: some relevant laboratory measurements. Mon.Not.R.Astron.Soc. 219, 89.

Lacy, J.H., Baas, F., Allamandola, L.J., Persson, S.E., McGregor, P.J., Lonsdale, C.J., Geballe, T.R. and van de Buit, C.E.P., 4.6 micron absorption features due to solid phase CO and cyanogroup molecules towards compact infra-red sources. Astrophys. J. 276, 533.

Langer, W.D., Graedel, T.E., Frerking, M.A. and Armentrout, P.B., 1984, Carbon and oxygen isotope fractionation in dense interstellar clouds. Astrophys. J. 277, 581.

Larson, H.P., Davis, D.S., Black, J.H. and Fink, U., 1985, Interstellar absorption features towards the compact infrared source W33A. Astrophys. J. 299, 873.

Léger, A. and d'Hendecourt, L., 1985, Are polycyclic aromatic hydrocarbons the carriers of the diffuse interstellar bands in the visible? Astron. Ap. 146, 81.

Léger, A., Jura, M. and Omont, A., 1985, Desorption from interstellar grains. Astron. Ap. 144, 147.

Léger, A. and Puget, J.L., 1984, Identification of the unidentified IR emission features of interstellar dust? Astron. Ap. 137, L5.

Lepp, S. and Dalgarno, A., 1987, Polycyclic aromatic hydrocarbons in interstellar chemistry. Astrophys. J. submitted.

Lepp, S., Dalgarno, A. and Sternberg, A., 1987, The abundance of H_3^+ in dense interstellar clouds. Astrophys. J. submitted.

Leung, C.M., Herbst, E. and Huebner, W.F., 1984, Synthesis of complex molecules in dense interstellar clouds via gas-phase chemistry: a pseudo time-dependent calculation. Astrophys. J. Suppl.56, 231.

Marquette, J.B., Rowe, B.R., Dupeyrat, G., Poissant, G. and Rebrion, C., 1985, Ion polar molecule reactions: a CRESU study of He^+, C^+, N^+ + H_2O, NH_3 at 27, 68 and 163K. Chem. Phys. Lett. 122, 431.

Millar, T.J., Adams, N.G., Smith, D. and Clary, D.C., 1985, The HCS^+/CS abundance ratio in interstellar clouds. Mon.Not.R.Astron.Soc. 216, 1025.

Millar, T.J. and Freeman, A., 1984, Chemical modelling of molecular sources. I. TMC-1. Mon.Not.Roy.Astron.Soc. 207, 405; II. L183 ibid 207, 425.

Millar, T.J., Leung, C.M. and Herbst, E., 1987, How abundant are complex interstellar molecules? Astron. Ap. in press.

Millar, T.J. and Nejad, L.A.M., 1985, Chemical modelling of molecular sources. IV. Time-dependent chemistry of dark clouds. Mon.Not. Roy.Astron.Soc. 217, 507.

Mitchell, G.F. and Huntress, W.T., 1979, Long chain carbon molecules and diffuse interstellar bands. Nature 278, 722.

Mitchell, G.F., Huntress, W.T. and Prasad, S.S., 1979, Interstellar synthesis of the cyanopolyynes and related molecules. Astrophys. J. 233, 102.

Omont, A., 1986, Physics and chemistry of interstellar polycyclic aromatic molecules. Astron. Ap. 164, 159.

Oppenheimer, M. and Dalgarno, A., 1975, The formation of carbon monoxide and the thermal balance in interstellar clouds. Astrophys. J. 200, 419.

Prasad, S.S. and Huntress, W.T., 1980b, A model for gas phase chemistry in interstellar clouds. II. Non-equilibrium effects and effects of temperature and activation energies. Astrophys. J. 239, 151.

Prasad, S.S. and Huntress, W.T., 1980a, A model for gas phase chemistry in interstellar clouds. 1. The basic model, library of chemical reactions and chemistry among C, N and O compounds. Astrophys. J. Suppl. 43, 1.

Prasad, S.S. and Tarafdar, S.P., 1983, UV radiation field inside dense clouds: its possible existence and chemical implications. Astrophys. J. 267, 403.

Smith, D., 1987, Interstellar molecules. Phil.Trans.Roy.Soc.Lond. in press.

Sternberg, A., Dalgarno, A. and Lepp, S., 1987, Cosmic ray induced photodestruction of interstellar molecules in dense clouds. Astrophys.J. submitted.

Su, T. and Bowers, M.I., 1979, Classical ion-molecule collision theory, p.84 in Gas phase ion chemistry (ed. M.T. Bowers) Academic, New York.

Takayanagi, K., 1978, Low energy ion-polar molecule collisions - the perturbed rotational state approach. J.Phys.Soc. Japan 45, 976.

Tarafdar, S.P., Prasad, S.S., Huntress, W.T., Villere, K.R. and Black, D.C., 1985, Chemistry in dynamically cooling clouds. Astrophys. J. 289, 220.

Van der Zwet, G.P. and Allamandola, L.J., 1985, Polycyclic aromatic hydrocarbons and the diffuse interstellar bands. Astron. Ap. 146, 76.

Watt, G.D., 1983, Time-dependent chemistry - 1. Modelling of a static cloud. Mon.Not.Roy.Astron.Soc. 205, 21.

Watt, G.D., 1985, Time-dependent chemistry - II. Dependence of the chemistry on the initial [C] - [O] ratio. Mon.Not.R.Astron.Soc. 212, 93.

Whittet, D.C.B., Longmore, A.J. and McFadzean, A.D., 1985, Solid CO in the Taurus dark clouds. Mon.Not.Roy.Astron.Soc. 216, 45P.

Williams, D.A., 1985, On the abundance of molecular hydrogen in the galaxy. Q.Jl.Roy.Astron.Soc. No.4, vol.26, 463.

Williams, D.A. and Hartquist, T.W., 1984, on C^o and CO in dense interstellar clouds - evidence that cloud material is frequently shocked. Mon. Not.Roy.Soc. 210, 141.

ANTARCTIC OZONE: A NEW CHALLENGE

Michael B. McElroy

Harvard University, Department of Earth and
Planetary Sciences
29 Oxford Street, Cambridge, MA 02138

1. INTRODUCTION

David Bates' research career extends over half a century, covers at least half a dozen fields, and continues to the present with undiminished vigor. An interest in the physics and chemistry of the atmosphere has been a persistent theme.

David made trail-blazing contributions to our understanding of atomic and molecular processes in the ionosphere. He pioneered early theoretical studies of the aurora and airglow. He pondered, before it was fashionable, factors influencing the composition of the atmosphere, not just short-lived photochemical products such as ozone, but also longer-lived gases such as carbon monoxide, methane, and nitrous oxide. He recognized the importance and subtlety of the ties that link the atmosphere to life and to the world of soil and ocean below. He foresaw the emerging importance of industrial man as an influence on the global environment. The seeds he planted and the flowers they grew are tribute to a remarkable and varied career.

David tackled large problems. He was particularly taken by issues that seemed beyond reach. In the absence of direct observational data, how could meaningful statements be made concerning the temperature of the outermost regions of the upper atmosphere? David's approach, with Holmesian skill, was to collect clues, nuggets invisible to others. He weighed the evidence and when time was right pronounced. The case was presented invariably with style. Watson, the reader, was inspired and illuminated. The paper presenting his conclusions concerning the temperature of the upper atmosphere (Bates, 1951) improves with age, like a classic of literature. A flower is grown; its beauty persists.

It would be easy to continue the horticultural metaphor, to present a guided tour of flowers planted by David Bates over the past fifty years. It seems inappropriate to do so here. David needs no introduction to the garden. He knows all its nooks and crannies. Better, if possible, to introduce the master gardener to a new specimen. New flowers, of course, are rare. I believe, however, that we can bring him at least a seedling, a plant with promise. We need his advice. Can he foretell its development by inspecting the embryo? Is our seedling coded to flower, or, perish the thought, is it destined to evolve the heartbreak of a weed?

Ozone has been disappearing over Antarctica during spring in recent years. The first report of this surprising development appeared only two years ago (Farman et al., 1985). The drop is not small. The column density of ozone over Halley Bay on the Antarctic coast (76°S) has fallen from about 320 Dobson units (DU)* in October 1975 to less than 200 DU in October 1984 (Farman et al., 1985). Concentrations observed over Antarctica in 1985 dropped below 150 DU (Stolarski, 1986), less than the value observed normally at the equator, close to the lowest value observed ever on the planet. We shall attempt to summarize what is known about the phenomenon in the following. There are some intriguing clues, the hint of a blossom about to burst; always, however, the peril that the flower may transform to a weed.

We begin, in Section 2, with a brief summary of processes regulating the abundance and global distribution of O_3 under normal unperturbed conditions, continuing in Section 3 with a more particular account for the Antarctic. Elements of a theory for the Antarctic phenomenon, the ozone hole as it has come to be known by the popular press, are presented in Section 4. Needs and prospects for new data are addressed in Section 5. Concluding remarks appear in Section 6.

2. CHEMISTRY AND DYNAMICS OF STRATOSPHERIC O_3

It is convenient to consider O and O_3 in combination, a family of compounds termed odd oxygen by Chapman (1930). Odd oxygen is produced by photodissociation of O_2, mainly in the optically forbidden Herzberg transition at wavelengths below 2400 Å

$$h\nu + O_2 \rightarrow O + O. \tag{1}$$

The atoms formed in (1) attach to O_2 forming O_3, through the three body reaction

$$O + O_2 + M \rightarrow O_3 + M. \tag{2}$$

The relative abundance of O and O_3 is set by (2) and by

$$h\nu + O_3 \rightarrow O + O_2. \tag{3}$$

In a pure oxygen system, odd oxygen is removed by

$$O + O_3 \rightarrow O_2 + O_2 \tag{4}$$

Most of the work in stratospheric chemistry since Chapman's pioneering paper in 1930 has been concerned with attempts to enhance the rate for (4).

The most important sinks for odd oxygen in the contemporary stratosphere, in addition to (4), are

$$\begin{aligned} NO + O_3 &\rightarrow NO_2 + O_2 \\ NO_2 + O &\rightarrow NO + O_2 \end{aligned} \tag{5}$$

* 1 DU is equal to 2.7×10^{16} molecules cm^{-2}

and

$$Cl + O_3 \rightarrow ClO + O_2$$
$$ClO + O \rightarrow Cl + O_2. \quad (6)$$

Reactions (5) and (6) are equivalent to (4). Catalysis, exemplified by (5) and (6), accounts for about 80% of the net sink for atmospheric odd oxygen.

The oxides of nitrogen in (5) are produced by decomposition of N_2O,

$$O(^1D) + N_2O \rightarrow NO + NO, \quad (7)$$

where $O(^1D)$ is produced by photolysis of O_3. The chlorine radicals in (6) are produced by decomposition of the industrial chlorocarbons $CFCl_3$, CF_2Cl_2, and CH_3CCl_3, with a small background contribution due to naturally produced CH_3Cl. Nitrogen radicals can be removed from the stratosphere, temporarily at least, by conversion to compounds such as N_2O_5 and HNO_3. The sink reactions in this case are

$$NO_2 + NO_3 + M \rightarrow N_2O_5 + M \quad (8)$$

and

$$OH + NO_2 + M \rightarrow HNO_3 + M. \quad (9)$$

Chlorine radicals are removed by reactions forming HCl, HOCl, and $ClNO_3$:

$$Cl + CH_4 \rightarrow HCl + CH_3, \quad (10)$$

$$ClO + HO_2 \rightarrow HOCl + O_2, \quad (11)$$

and

$$ClO + NO_2 + M \rightarrow ClNO_3 + M. \quad (12)$$

The compounds N_2O_5, HNO_3, HCl and $ClNO_3$ provide temporary reservoirs for nitrogen, chlorine, and hydrogen radicals. Radicals are recycled by photolysis and by reactions such as

$$N_2O_5 + M \rightarrow NO_2 + NO_3 + M, \quad (13)$$

$$OH + HNO_3 \rightarrow H_2O + NO_3, \quad (14)$$

and

$$OH + HCl \rightarrow H_2O + Cl. \quad (15)$$

The sink reactions (8)-(12) increase generally in efficiency at low altitude. As a consequence the abundance of radicals is relatively small below about 30 km. The abundance of Cl + ClO is sensitive to the abundance of NO_2 at low altitudes since (12) is the dominant sink for chlorine radicals.

The essential elements of stratospheric chemistry can be elucidated using the simple, pure oxygen, model presented by Chapman (1930). Odd oxygen is produced by (1), removed by (4). In steady state,

$$J_1[O_2] = k_4[O][O_3] \quad (16)$$

where J_1 (sec^{-1}) and k_4 (cm^3sec^{-1}) denote rate constants for (1) and (4), respectively, and the number density of i (cm^{-3}) is given by [i]. The rate for production of O by (1) is small compared to that by (3) for two reasons. First, radiation with sufficient energy to dissociate O_2 is absorbed at relatively high levels in the atmosphere, not just by O_2 but also by O_3. Second, the cross-section for (1) is less than that for (3) by almost a factor of 10^7. The density of O_2, though greater than that by O_3 by more than a factor of 10^6, is insufficient to compensate for this deficit. The concentration of O is determined thus by a balance of (2) and (3):

$$O = \frac{J_3[O_3]}{k_2[O_2][M]}, \qquad (17)$$

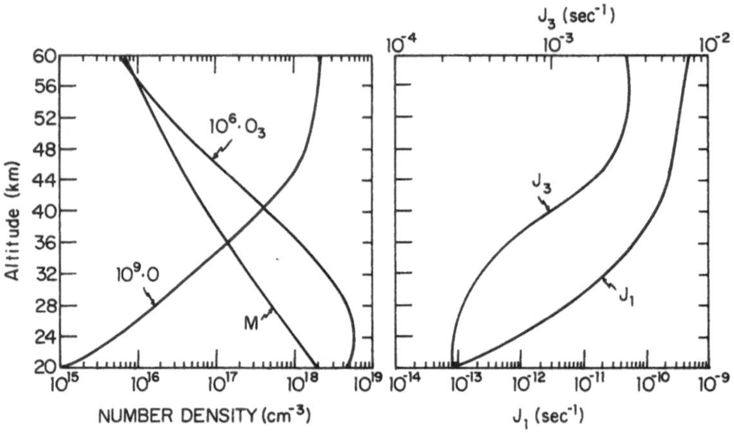

Figure 1. Height profiles calculated for O_3, O and M, using (17) and (18), illustrating height variations of the rates for photolysis, J_1 and J_3.

where J_3 (sec^{-1}) is the rate for photolysis of O_3 and k_2 (cm^6sec^{-1}) is the rate constant for (2). Substituting for O in (16) gives

$$[O_3] = \left\{ \frac{J_1k_2[O_2]^2[M]}{J_3k_4} \right\}^{1/2} \qquad (18)$$

Rate constants are more or less invariant with height at high altitude. The relative abundance of O_2 is essentially independent of altitude, at least for altitudes below about 80 km. It follows that the concentration of O_3 should fall off with altitude, at high altitude, as $[M]^{3/2}$. At low altitude, J_1 is vanishingly small. The concentration of O_3 as given by (18) assumes a maximum at about 25 km, while the concentration of O increases with increasing altitude, as $[M]^{-1/2}$. The general behavior of $[O_3]$ and [O], as given by (17) and (18), is illustrated in Figure 1.

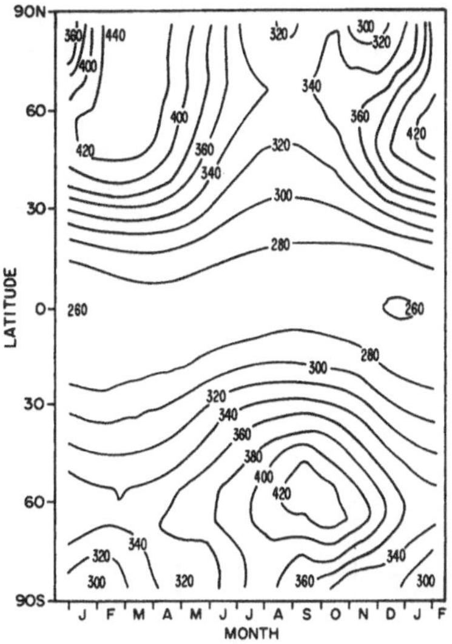

Figure 2. Column abundance of O_3 (DU) as functions of latitude and time of year. (From Atmospheric Ozone 1985.) World Meteorological Organization Global Ozone Research and Monitoring Project Report No. 16.

Chapman's scheme provides a remarkably accurate representation of O_3 from about 30 to 80 km. Concentrations of O_3 computed on the basis of (1)-(4) agree with observation to within about a factor of 3 at low and mid latitudes. The model is deficient at high altitude where it fails to allow for transport and for removal of O by

$$O + O + M \rightarrow O_2 + M. \quad (19)$$

It must be adjusted also at low altitude where the chemical lifetime of O_3 is long, and where the distribution of the gas is determined largely by atmospheric motion. Ozone tends to accumulate in regions where it is protected from chemical loss, and this accounts for the gross features of the latitudinal distribution shown in Figure 2. Largest column densities are observed at high latitudes where removal is least efficient. The data in Figure 2 suggest that at high latitudes in the southern hemisphere O_3 should vary from a low of about 300 DU in local summer to a high of about 385 DU in late winter. There is no indication in Figure 2 of the dramatic drop in O_3 observed in spring over the Antarctic since 1975.

3. OZONE OVER ANTARCTICA

The low concentrations of O_3 observed in recent years over Antarctica in spring are associated with a circulation pattern known as the polar vortex. Temperatures are exceedingly low in winter over Antarctica. A large gradient develops in temperature from south to north, associated in the polar region with a strong zonal flow. Air within the vortex is effectively isolated until the vortex breaks up, typically in mid October. The vortex covers a relatively large area, extending from the pole to latitudes of about 60°S. The decline in O_3 begins, as best one can tell, in late August or early September (Chubachi, 1984; Mount et al., 1987). The region of low O_3 covered an area of 18.8 million km^2 in October 1982, approximately twice the size of the United States (Schoeberl et al., 1986).

A theory for Antarctica O_3 must account first for the secular trend observed over the past decade. The data of Farman et al. (1985), representing a mean of observations during October at Halley Bay, are displayed in Figure 3. The satellite data (Stolarski et al., 1986) confirm the trend shown in Figure 3 and provide a wealth of additional detail on the morphology of the phenomenon. They indicate that the drop in O_3 in September 1983 averaged about 0.6% per day (Stolarski et al., 1986).

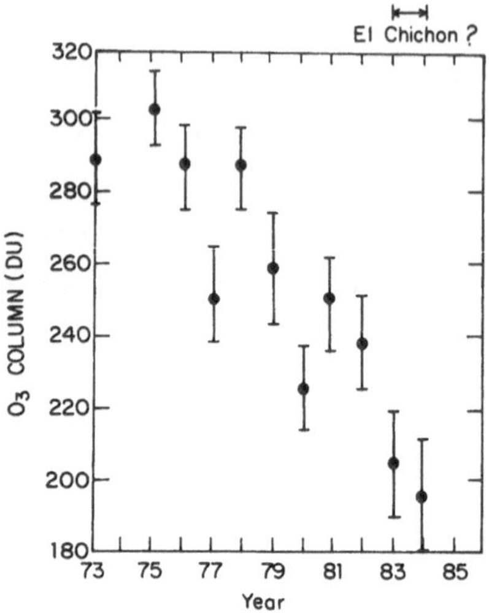

Figure 3. Monthly mean values of O_3 observed for October at Halley Bay 76°. From Farman et al., 1985.

The reduction in O_3 is confined for the most part to altitudes below 20 km. Several profiles adapted from ozonesonde data taken at Syowa (69°S) (Chubachi, 1984; Ozone data for the world, 1985) are shown in Figure 4. Curve (a) illustrates the mean of data for October for years prior to 1975. Curve (b) gives an average of profiles measured during the winter of 1982. Curves (c)-(e) show individual profiles obtained during the months of October in 1982 and 1983. Largest reductions in O_3 were observed at Syowa between about 12 and 22 km. Similar behavior was seen more recently at McMurdo station (78°S). Measurements by Hofmann et al. (1987) in September and October of 1986 suggest that loss of O_3 may be patchy. Layers 2 to 3 km thick were observed where O_3 was effectively absent (depleted by more than 75%), interspersed with regions which seemed relatively unperturbed. Loss of O_3 appeared to begin and end early at the higher altitudes, 18-20 km, commencing later and ending later at lower altitudes, 12-14 km. Removal of O_3 was observed throughout the observing period, late August to late October, at intermediate altitudes.

The disappearance of O_3 appears to be correlated with the evaporation of polar stratospheric clouds (PSCs) (Hamill et al., 1986). It is unclear whether this reflects merely a coincidence in time or a more fundamental underlying relationship. PSCs are observed most frequently when temperatures drop below about 190°K (McCormick et al., 1982). They are thought to be composed mainly of water ice (Steele et al., 1983; Kent et al., 1986). Toon et al. (1986), however, raised the possibility that they might be formed predominantly of hydrated nitric and hydrochloric acids. We return to this possibility later in connection with the discussion of possible theories for the Antarctic hole.

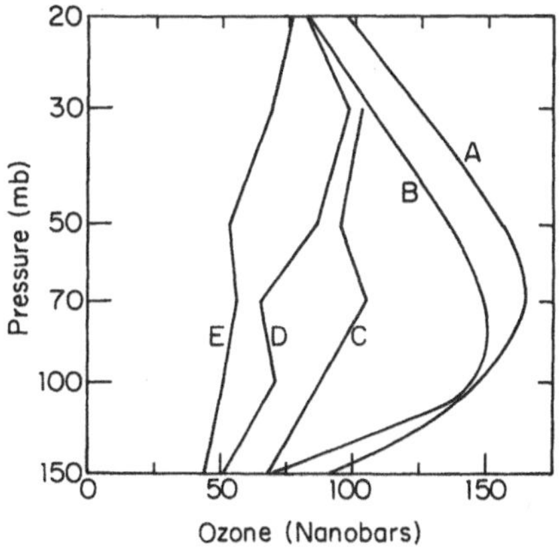

Figure 4. Vertical distributions of ozone at Syowa (69°, 40°E). Curve (a) gives an average of 26 soundings made between 1968 and 1974, in October. The average column density recorded during this period was 337 DU. Curve (b) reflects an average of soundings made in June, July and August, 1982; the average ozone column was 320 DU. Curves (c), (d) and (e) summarize measurements for 23 September, 1982, 14 October, 1982, and 17 October, 1983; the ozone columns were 256, 241 and 217, respectively. The mean ozone column for October 1982 was 236 DU, 260 DU for October 1983. From McElroy et al., 1986a.

4. THEORETICAL CONSIDERATIONS

Most telling, perhaps, of the observational constraints noted above is the information on the height profile. It is difficult to devise a mechanism to account for loss of O_3 of the magnitude observed, even more challenging to find a scheme that operates mainly below 20 km. The reductions in O_3 observed by Hofmann et al. (1987) near 18 km imply rates for removal of odd oxygen in early September of about 10^6 cm^{-3} sec^{-1}, a half life of about 25 days. To put matters in context, the abundance of O as calculated for 18 km at 80°S on September 15 is only $3x10^5$ cm^{-3}: removal of odd oxygen at this altitude by (4) takes place at a rate of $1.4x10^2$ cm^{-3} sec^{-1}, which implies a lifetime of nearly 1000 years for O_3. The catalytic sequences (5) and (6) are potentially more efficient: they also fail, however, by at least a factor of 10, even with the most optimistic assumptions concerning the source of ClO and NO_2. Reactions (4)-(6) are limited at low altitude ultimately by the supply of O.

The reactions

$$OH + O_3 \rightarrow HO_2 + O_2$$

$$HO_2 + O_3 \rightarrow OH + 2O_2, \qquad (20)$$

$$Cl + O_2 \rightarrow ClO + O_2$$
$$HO_2 + ClO \rightarrow HOCl + O_2$$
$$h\nu + HOCl \rightarrow OH + Cl$$
$$OH + O_3 \rightarrow HO_2 + O_2, \qquad (21)$$

$$ClO + ClO + M \rightarrow Cl_2O_2 + M$$
$$Cl_2O_2 + h\nu \rightarrow Cl + ClO_2$$
$$ClO_2 + M \rightarrow Cl + O_2 + M$$
$$2[Cl + O_3 \rightarrow ClO + O_2], \qquad (22)$$

and

$$Cl + O_3 \rightarrow ClO + O_2$$
$$Br + O_3 \rightarrow BrO + O_2$$
$$ClO + BrO \rightarrow Cl + Br + O_2, \qquad (23)$$

provide additional paths for removal of odd oxygen. Supply of O poses no problem in this case: reactions (20)-(23) are equivalent to:

$$O_3 + O_3 \rightarrow 3O_2 \qquad (24)$$

Solomon et al. (1986) suggested that (21) could play an important role for O_3 in Antarctica. McElroy et al. (1986a) favored (23), while (22) was introduced by Molina and Molina (1987).

It is easy to show that the rate for (21) is insufficient to account for the decline in O_3 observed by Stolarski et al. (1986) and Hofmann et al. (1987), except under extraordinary circumstances. The rate for photolysis of HOCl is about 4.5×10^{-5} sec^{-1} under conditions appropriate for the Antárctic in September. A loss rate of 3.0×10^{5} cm^{-3} sec^{-1}, equivalent to a drop in O_3 of 0.6% per day, would require a concentration of HOCl of 1.1 parts per billion (ppb) if photolysis HOCL were to represent the rate limiting step for removal of O_3 by (21). The concentration of total chlorine is presently about 2.6 ppb. Current models suggest that HOCl should account for no more than 2% of stratospheric Cl_x, defined as HCl + $ClNO_3$ + HOCl + ClO +Cl_2O_2. A major role for (21) would require almost complete conversion of chlorine to ClO + HOCl. This would imply low concentrations of NO_2 and concentrations of OH and HO_2 larger than values calculated in most contemporary models (McElroy et al., 1986a,b; Tung et al., 1986) by at least a factor of 100. Reactions (21), as discussed by Crutzen and Arnold (1986), could be important if the abundance of NO_x, defined as HNO_2 + HNO_3 + HNO_4 + NO + NO_2 + NO_3 + N_2O_5 + $ClNO_3$, was essentially zero. We find, with more realistic estimates for NO_x that (21) makes but a small contribution to the observed loss of O_3.

The atmosphere contains a variety of brominated organic compounds, some natural, others products of modern industrial activity. The constituent bromine atom concentration ranges from about 25 to 30 parts per trillion (ppt) (Singh et al., 1983; Berg et al., 1983; Rasmussen and Khalil, 1984). The most abundant of the brominated gases capable of reaching the stratosphere, between 30 and 70% of the total, is CH_3Br. Approximately half of this is natural, produced by methylating organisms in the sea (Singh et al., 1983); the balance is industrial, reflecting in part, applications of CH_3Br as an agricultural fumigant. Concentrations of the long-lived compounds Halon 1301 (CF_3Br) and Halon 1211 (CF_2ClBr) are increasing rapidly at the present epoch, reflecting expanding use of these products as fire extinguishers. Lifetimes of the Halons are long, 110 and 25 years respectively (Molina et al., 1982). Levels may be expected to rise steadily in the coming years, leading to about a 5 fold increase in the associated source of stratospheric bromine even if emission rates stabilize at values applicable today. An increase in emissions by a factor of 6 seems not unreasonable and would provide, in steady state, an abundance of stratospheric bromine as high as 100 ppt (Prather et al., 1984).

A concentration of 20 ppt BrO, in combination with 200 ppt ClO, with rate constants for reactions (23) recommended by DeMore et al. (1985) would imply a rate for removal of O_3 at 18 km of 2.0×10^5 cm^{-3} sec^{-1}, about enough to account for removal of O_3 observed in September. Loss by (22) is comparable, about 1×10^5 cm^{-3} sec^{-1}, if the rate constant for

$$ClO + ClO + M \rightarrow Cl_2O_2 + M, \qquad (25)$$

is taken equal to 1.7×10^{-31} cm^{+6} sec^{-1} (Cox, 1986; Molina and Molina, 1987). The rate for removal of O_3 by (22) varies as the square of the concentration of ClO, for moderate levels of ClO. At high concentrations of ClO, the dimer, Cl_2O_2, is a significant reservoir for reactive chlorine and effects of (22) tend to saturate. With current understanding of relevant kinetics, a combination of (22) and (23) could account for a significant fraction of O_3 loss observed below 20 km. In this case, at least 10% of the available supply of chlorine must be converted to ClO. A necessary but not sufficient condition for this to hold is that the abundance of NO_2 be less than that of ClO; otherwise an unacceptably large fraction of reactive chlorine will tend to end up as $ClNO_3$.

Toon et al. (1986) suggested that the cold temperatures of the Antarctic stratosphere could promote formation of solid phases containing significant quantities of HNO_3. They argued that condensation of HNO_3 could provide the low NO_x environment needed for reactions (21)-(23), and suggested that PSCs might be composed mainly of concentrated solutions of nitric acid, with lesser quantities of HCl and H_2SO_4. They assumed, in their analysis, that solid phases formed by freezing mixtures of H_2O with HNO_3, and H_2O with HCl, have the same composition as the liquid form which they precipitate. Failure to account for the specific crystalline phases leads to erroneous conclusions in the case of HCl-H_2O: an analysis allowing for the identity of crystalline phases (McElroy et al., 1986b; Salawitch et al., 1987) suggests that vapor pressures of HCl in the stratosphere are too low by at least a factor of 10 to permit condensation of hydrated HCl. Similar analyses for HNO_3-H_2O (McElroy et al., 1986b; Crutzen and Arnold, 1986) suggest that, despite inaccuracies in their treatment of the relevant thermodynamics, Toon et al. (1986) may be correct in their assertion that condensation could provide a significant sink for stratospheric HNO_3. The potential importance of HNO_3 condensation as a sink for NO_x is illustrated in Figure 5. The curve labelled "profile" illustrates concentrations of HNO_3 expected in the absence of condensation. The upper horizontal axis gives vapor pressures of HNO_3 calculated assuming equilibrium with solid crystalline monohydrate. Corresponding temperatures are shown on the lower horizontal axis. Curve A presents a temperature profile reported by McCormick et al. (1982) for 1 September, 1979. With curve A, we expect HNO_3 to condense between 14.5 and 20.5 km. If the temperature were lowered by 5°K, curve B, depletion of HNO_3 would be larger and would extend from about 10 to 22 km.

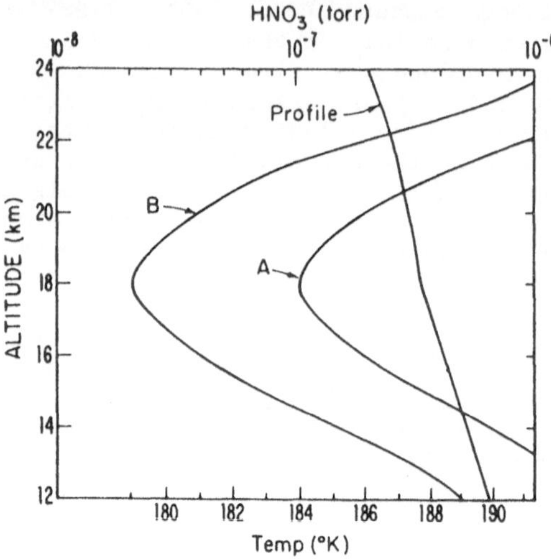

Figure 5. Vertical profile of HNO_3 over Antarctica, from Ko et al. (1984), is compared with saturation vapor pressures calculated for $HNO_3 \cdot H_2O$ crystals. The temperature profile A is as given by McCormick et al. (1982) for 1 September, 1979. Profile B reflects a reduction in temperatures by 5°K with respect to A. From McElroy et al., 1986b.

Toon et al.'s (1986) proposal that HNO_3 should be the dominant component of PSCs is less plausible than the simple suggestion that the gas might condense in the stratosphere. They argued that the opacity of PSCs would be much larger than values observed, typically less than 10^{-2} km^{-1}, if PSCs contained appreciable quantities of H_2O. It is difficult, however, to escape the conclusion that H_2O must condense as well as HNO_3: the freezing temperature of H_2O differs from that of HNO_3 hydrates by no more than a few degrees, and the abundance of H_2O exceeds that of HNO_3 by about a factor of 500. There may be a simple resolution to the dilemma. It is likely, particularly at higher altitudes, that the condensing particles will grow to a size large enough to fall. This provides a mechanism for vertical transport of HNO_3 and H_2O, and offers, potentially, an explanation for the vertical structure observed in O_3. Regions that have lost HNO_3 may be poised to lose O_3, assuming that residual concentrations of chlorine and bromine are high enough to permit an appreciable role for (21)-(23). Condensation can serve to condition the atmosphere, paving the way for later loss of O_3. It is obvious that this conditioning will be localized and that it will be very sensitive to temperature. Appreciable loss of O_3 may require the coincidence of two factors, low temperatures and high concentrations of reactive halogens.

Hamill et al. (1986) showed evidence that the drop in O_3 occurs coincident with evaporation of PSCs near 18 km. It is reasonable to suppose that the condensation sequence, on cooling, involves first incorporation of H_2O in a nucleus of concentrated H_2SO_4, then condensation of HNO_3, and finally, addition of large quantities of H_2O. The sequence would repeat in reverse order as the atmosphere began to warm up in spring. If we assume that HNO_3 is essentially absent from the gas phase in early September in regions where loss of O_3 is significant, it seems unreasonable to suppose that the concentration of HNO_3 should remain low until evaporation of H_2O is nearly complete. Removal of O_3 would terminate in the late stages of evaporation, as temperatures rose above 190°K. Nitric acid would return to the gas phase. Photochemical reactions would result in production of NO_2 and associated immobilization of chlorine and bromine radicals.

Results of McElroy et al. (1986b), illustrating the behavior of spring time Antarctic O_3, with different assumptions concerning the abundance of HNO_3 and chlorine radicals, are shown in Figure 6. The vertical profile of Cl_x was constructed as described by McElroy et al. (1986a). A one dimensional model was used to define Cl_x at 30°S. Results were scaled among mixing surfaces to estimate Cl_x profiles at 45, 60, and 80°S. Abundances of ClO_x, defined as $ClO + ClNO_3 + HOCl + OClO + ClOO + Cl_2O_2$, were computed for each of the four latitudes for a solar declination of 0°. The abundance of ClO_x, expressed as a ratio relative to Cl_x, was averaged along the mixing surfaces to estimate concentrations of ClO_x incorporated in the polar vortex.

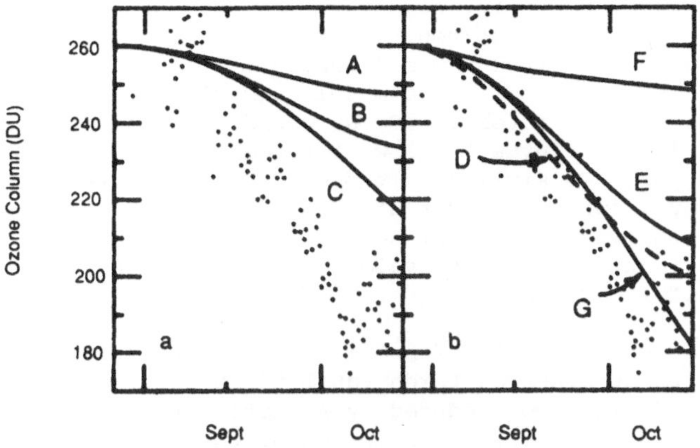

Figure 6. Depletion of total ozone (DU) during September and October. Panel (a) gives model results ignoring effects of Br radicals: curve A, HNO_3 reduced to 10% of unperturbed, conventional chlorine radical chemistry without ClO dimers; curve B, same HNO_3 as A, with ClO dimers; curve C, HNO_3 reduced to 2% of unperturbed, with ClO dimers.
Panel (b) gives results for models including Br radicals: curve D, HNO_3 10% of unperturbed, no ClO dimers; curve E, same as D with ClO dimers; curve F, same as E, with unperturbed HNO_3; curve G, same as E with HNO_3 2% of unperturbed. Data points are from Stolarski et al. (1986). From McElroy, 1986b.

McElroy et al. (1986a,b) assumed that HOCl was the dominant form of ClO_x as the Antarctic emerged from winter, with NO_x present mainly as HNO_3. They supposed that $ClNO_3$ was converted to HOCl and HNO_3 over the course of the polar night. McElroy et al. (1986a) and Solomon et al. (1986) hypothesized that this could occur on the surface of polar aerosols. Two possible reactions were suggested,

$$ClNO_3 + H_2O \rightarrow HOCl + HNO_3, \qquad (26)$$

and

$$ClNO_3 + HCl \rightarrow Cl_2 + HNO_3. \qquad (27)$$

McElroy et al. (1986a) considered (27) less likely than (26), in that it requires reaction of two minor species on a surface presumed to be composed mainly of H_2O. Recent experiments by M.J. Molina (private communication, 1987) suggest that HCl may adsorb to ice surfaces, that it may dissolve in ice at low temperatures. This could mitigate significantly the objections to (27). Reaction (27) does double duty, not only removing reactive nitrogen but also enhancing the supply of reactive chlorine. It could allow a larger initial abundance of reactive chlorine than that considered in Figure 6. For purposes of this calculation, effects of heterogeneous chemistry were assumed unimportant after September 1. It matters little whether reactive chlorine is present initially as HOCl or as Cl_2: it is converted in either case rapidly to ClO.

The results in Figure 6a explore possible effects of chlorine chemistry, reactions (21) and (22), on removal of O_3. Results are shown for a latitude of 80°S, beginning August 27 when the sun first rises above the horizon. The initial column density of O_3 was set equal to 260 DU, in agreement with data from Syowa (Chubachi, 1984). The dimer chemistry, reaction (22), was omitted in the model summarized by curve A. Nitric acid was reduced by a factor of 10, as would be expected if temperatures dropped below 185 K. Curve B allows for removal of O_3 by (22), with concentrations of HNO_3 similar to those in A. Concentrations of HNO_3 were reduced by a further factor of 5 in model C. Rates of O_3 loss average 0.3, 0.6, and 0.8 DU per day in September in models A-C respectively, significantly less than values observed since 1982 which average typically about 1.6 DU per day.

Figure 6b shows effects of bromine, reactions (23). The concentration of reactive bromine was taken equal to 25 ppt. Curve D illustrates results obtained with HNO_3 reduced by a factor of 10, ignoring effects of the dimer chemistry, (22). Inclusion of the dimer chemistry lowers the concentration of ClO and results in a smaller net loss of O_3, as shown by curve E. With unperturbed HNO_3, loss of O_3 is only 0.3 DU per day, curve F, in contrast to the large and persistent loss obtained when HNO_3 was reduced by a factor of 50, as shown by curve G.

There are a number of possible ways to test the validity of the model described above. Most direct, of course, would be a measurement of elevated levels of ClO and BrO below 20 km. This would establish that conditions were set for catalytic removal of O_3. It would be important to understand then why concentrations of ClO and BrO were high. A step in this direction would be to check that NO_2 was low coincident with high ClO and BrO. If significant loss of O_3 was observed in the presence of high ClO and BrO, and low NO_2 and HNO_3, one would be satisfied that the theory was basically correct. It would remain to elaborate the role of heterogeneous chemistry, to define the influence of PSCs, and, most important perhaps, to identify the extent to which the ozone hole phenomenon might depend on the occurrence of unusually cold temperatures over the Antarctic in recent years. We note, in this context, evidence presented by Chubachi (1986) for significant cooling in spring at stratospheric altitudes over Syowa during the interval 1974-1985. This issue merits further attention, to determine whether the temperature trend is real, and if so, to define its geographic and temporal extent, to investigate by what connection, if any, it might be linked to the observed change in O_3.

Tung et al. (1986) pointed out that there are several possible paths for reaction of ClO and BrO. In addition to the step involved in the recombination scheme (23),

$$ClO + BrO \rightarrow Cl + Br + O_2, \tag{28}$$

there is a path that can lead to production of OClO,

$$ClO + BrO \rightarrow OClO + Br. \quad (29)$$

If (23) was important, we would expect significant production of OClO at night, as illustrated in Figure 7. Efficient photolysis of OClO, yielding ClO and O, would ensure rapid replacement of OClO by ClO following sunrise. We suggested, with some trepidation, that OClO in amounts required by the ClO-BrO theory should be readily detectable from the ground, using techniques pioneered by Noxon (1975). The molecule has a number of strong, characteristic features in the near ultraviolet portion of the spectrum which should permit its detection either in directly transmitted moonlight or in sunlight scattered over long pathlengths at twilight. The test was carried out in September 1986. A preliminary paper (Solomon et al., 1987) has appeared, announcing detection of OClO

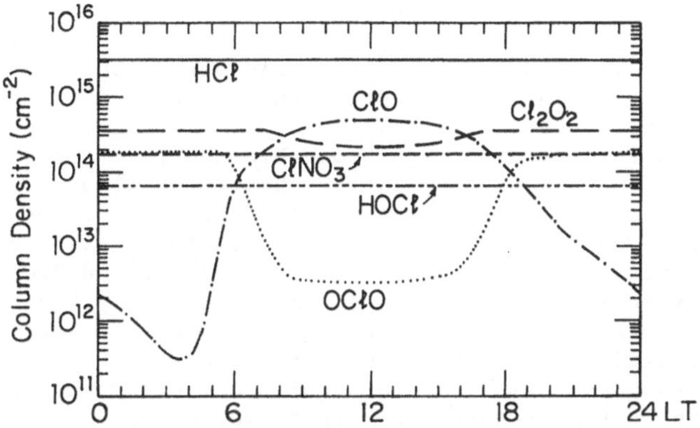

Figure 7. Densities for major chlorine species as functions of local time (LT) from midnight (0) integrated over altitude from 13 to 20 km of September 15. Results were obtained using the model described by McElroy et al. (1986b) with assumptions corresponding to those in Figure 6b, curve E.

in roughly the amounts required by the ClO-Bro theory (McElroy et al., 1986a; Rodriguez et al., 1986). High concentrations of OClO were found in August and September at McMurdo, associated with low concentrations of NO_2 and cold temperatures. Concentrations of OClO dropped below the detection threshold of the instrument around October 1. A search for OClO at other latitudes, notably at 40°N (Boulder, Colorado) and at 50°N (Gimli, Canada), has so far proved negative. In that reaction (29) is the only plausible source for OClO suggested to date, it seems reasonable to interpret the data from the Antarctic as evidence for elevated levels of ClO and BrO. It is difficult to escape the conclusion that the chemical environment of the springtime Antarctic stratoshpere is unusual. There is at least an a priori case to conclude that chlorine and bromine radicals are involved in springtime removal of O_3.

5. CONCLUDING REMARKS

The drop in O_3 observed over Antarctica in spring in recent years poses a number of important questions. The issues extend beyond the normal bounds of scientific enquiry, reaching squarely into the domain of the political. The politician demands straightforward answers. The questions are to the point and equivocation is greeted with impatience. Is the observed change in O_3 a harbinger of what lies ahead for the northern hemisphere? Can it be expected to spread to other times of the year and to more populated latitudes? The honest answer - we do not know - is not popular. It is important, though, that the distinction between fact and opinion be clear and explicit. It should be apparent when the scientist speaks as expert and when the answer reflects merely opinion. The politician should know what we know, and how and when we might expect to know more.

Before we begin to extrapolate our knowledge of the Antarctic to other times and places, it is essential that we develop a comprehensive picture of the physical and chemical processes responsible for removal of O_3 over Antarctica. This requires a model. Verification of the model demands quantitative predictions that can be subjected to observational scrutiny. A candidate model was outlined above. A quantitative prediction has been verified - the Antarctic stratosphere contains concentrations of OClO in rough agreement with expectation. The ultimate significance of the observation, however, is still unclear. For example, if (29) were the dominant path for reaction of ClO and BrO under stratospheric conditions, the observation of OClO would say little about removal of O_3. It would provide merely a constraint on the product of the abundances of ClO and BrO. To relate the OClO observation more directly to the stratosphere, there is a need for laboratory studies of the reaction of ClO and BrO. We need to define rates for individual reaction channels, to study their behaviour as a function of temperature. There is a general need to extend the range of temperatures available in the laboratory since extrapolation of rate data from high temperatures could lead to significant error when applied to the stratosphere, where temperatures may dip as low as 180°K. There is a particular need for studies of reactions involving production and removal of the ClO dimer. It would seem prudent at the same time to examine other species that might be expected to form under the distinctive environment of the Antarctic stratosphere, ClO·BrO for example.

Most of all, there is a need for additional observational data, in particular for in situ measurements of composition between about 12 and 20 km. What is the chemical environment in mid August, immediately prior to the return of sunlight following the long polar winter? How heterogeneous is the distribution of NO_x? What are the dominant forms of ClO_x? To what extent, as suggested above, is removal of NO_x the key in conditioning the Antarctic stratosphere for rapid springtime removal of O_3? How, in detail, does removal of O_3 relate to ambient temperature? Is the recent drop in temperature part of a natural fluctuation, or, more bothersome, could it signal the onset of a more persistent trend, related, for example, to increases in concentrations of gases such as CO_2, CH_4, N_2O, and the chlorofluorocarbons? Increases in the concentrations of these species are expected to cause an increase in the surface temperature of the Earth, with compensatory cooling of the stratosphere (Hansen et al., 1981). Is it possible that feedbacks involving O_3 and PSCs could exaggerate the impact in the stratosphere during Antarctic spring? Questions abound: answers are few. Such, however, is the challenge of science. As David Bates realized early in his career, if the answers are easy, the questions are often trivial.

How then do we deal with the politician? Is the loss of O_3 over

Antarctica transitory? Do we expect the hole to spread to other latitudes? Reductions in O_3 over Antarctica in spring will persist, in my opinion. I expect the phenomenon, though, to be episodic, with no clear trend. The severity of the drop will depend, I believe, on temperature. If temperatures between 10 and 20 km are low in early September, levels of O_3 may be expected to plunge. An extension to other latitudes is unlikely. The circumstances in Antarctica are unique, in my opinion, and implications for other environments are minimal. All this, I stress, is a matter of opinion. I reserve the right to change my mind.

ACKNOWLEDGEMENTS

I am indebted to D. Hofmann and S. Solomon for preprints of papers describing results from the expedition to McMurdo Station in 1986. Discussions with R. Salawitch were stimulating and always instructive. This work was supported by NASA grant NAG5-786 and by NSF grant ATM-84-13153 to Harvard University.

REFERENCES

Bates, D.R., 1951, The temperature of the upper atmosphere, Proc.Phys. Soc. (London), B64:805.

Berg, W.W., Sperry, P.D., Rahan, K.A. and Gladney, E.S., 1983, Atmospheric bromine in the Arctic, J. Geophys. Res., 88:6719.

Chapman, S., 1930, A theory of upper atmospheric ozone, Mem. R. Met. Soc. 3:103.

Chubachi, S., 1984, Preliminary result of ozone observation at Syowa Station from February 1982 to January 1983, Mem. Natn. Inst. Pol. Res. (Tokyo), Special Issue 34:13.

Chubachi, S., 1986, On the cooling of stratospheric temperature at Syowa, Antarctica, Geophys. Res. Lett., 13:1221.

Cox, R.A., 1986, private communication.

Crutzen, P.J. and Arnold, F., 1986, Odd nitrogen incorporation in polar stratospheric clouds: a possible cause for the springtime ozone decay in Antarctica, Nature, 651.

DeMore, W.B., Margitan, J.J., Molina, M.J., Watson, R.T., Golden, D.M., Hampson, R.F., Kurylo, M.J., Howard, C.J. and Ravishankara, A.R., 1985, Chemical kinetics and photochemical data for use in stratospheric modelling, Evaluation No. 7, JPL Publication 85-37, Jet Propulsion Laboratory, Pasadena, CA, 226 p.

Farman, J.C., Gardiner, B.G. and Shanklin, J.D., 1985, Large losses of total ozone in Antarctica reveal seasonal ClO_x/NO_x interaction, Nature, 315:207.

Hamill, P., Toon, O.B. and Turco, R.P., 1986, Characteristics of polar stratospheric clouds during the formation of the Antarctic ozone hole, Geophys. Res. Lett., 13:1288.

Hansen, J., Johnson, D., Lacis, A., Lebedeff, S., Lee, P., Rind, D. and Russell, G., 1981, Climate impact of increasing atmospheric carbon dioxide, Science, 213:957.

Hofmann, D.J., Harder, J.W., Rolf, S.R. and Rosen, J.M., 1987, Balloon-borne observations of the development and vertical structure of the Antarctic ozone hole in 1986, Nature, 326:59.

Kent, G.S., Poole, L.R. and McCormick, M.P., 1986, Characteristics of arctic polar stratospheric clouds as measured by airborne lidar, J. Atmos. Sci., 43:2149.

McCormick, M.P. and Trepte, C.R., 1982, Polar stratospheric cloud sightings by SAM II, J. Atmos. Sci., 39:1387.

McElroy, M.B., Salawitch, R.J., Wofsy, S.C. and Logan, J., 1986, Reductions of antarctic ozone to synergistic interactions of chlorine and bromine, Nature, 321:759.
McElroy, M.B., Salawitch, R.J. and Wofsy, S.C., 1986b, Antarctic O_3: Chemical mechanisms for the spring decrease, Geophys. Res. Lett., 13:1296.
Molina, L.T., Molina, M.J. and Rowland, F.S., 1982, Ultraviolet absorption cross sections of several brominated methanes and ethanes of atmospheric interest, J. Phys. Chem., 86:2672.
Molina, L.T. and Molina, M.J., 1986, Production of Cl_2O_2 from the self-reaction of the ClO radical, J. Phys. Chem., 433.
Molina, M.J., 1987, private communication.
Mount, G.H., Sanders, R.W., Schmeltekopf, A.L. and Solomon, S., 1986, Visible spectroscopy at McMurdo Station, Antarctica. 1. Overview and daily variations of NO_2 and O_3 during austral spring, 1986. (submitted to J. Geophys. Rev., 1987).
Noxon, J.F., 1975, Science, 189:547.
Ozone Data for the World, Indexes 6, 14 and 19 (Atmospheric Environmental Service, Toronto, Ontario, Canada, 1985).
Prather, M.J., McElroy, M.B. and Wofsy, S.C., 1984, Reductions in ozone at high concentrations of stratospheric halogens, Nature, 312:227.
Rasmussen, R.A. and Khalil, M.A.K., 1984, Gaseous bromine in the arctic haze, Geophys. Rev. Lett., 11:433.
Rodriguez, J.M., Ko, M.K.W. and Sze, N.D., 1986, Chlorine chemistry in the Antarctic stratosphere: impact of OClO and Cl_2O_2 and implications for observations, Geophys. Res. Lett., 13:1291.
Salawitch, R.J., Wofsy, S.C. and McElroy, M.B., 1987, Will HCl condense in the Antarctic stratosphere? (to be published).
Schoeberl, M.R., Krueger, A.J. and Newman, P.A., 1986, The morphology of Antarctic total ozone as seen by TOMS, Geophys. Res. Lett., 13:1217.
Singh, H.B., Salas, L.J. and Stiles, R.E., 1983, Methyl halides in and over the eastern Pacific (40^oN-32^oS), J. Geophys. Res., 88:3684.
Solomon, S., et al., 1986, On the depletion of Antarctic ozone, Nature, 321:755.
Solomon, S., Mount, G.H., Sanders, R.W. and Schmeltekopf, A.L., 1987, Visible spectroscopy at McMurdo Station, Antarctica (submitted to J. Geophys. Res.).
Steele, H.M., Hamill, P., McCormick, M.P. and Swissler, T.J., 1983, The formation of polar stratospheric clouds, J. Atmos. Sci., 40:2055.
Stolarski, R.S., Krueger, A.J., Schoeberl, M.R., McPeters, R.D., Newman, P.A. and Alpert, J.C., 1986, Nimbus 7 satellite measurements of the springtime Antarctic ozone decrease, Nature, 322:808.
Toon, O.B., Hamill, P., Turco, R.P. and Pinto, J., 1986, Condensation of HNO_3 and HCl in the winter polar stratospheres, Geophys. Res. Lett., 13:1284.
Tung, K.K., Ko, M.K.W., Rodriguez, J.M. and Sze, N.D., 1986, Are Antarctic ozone variations a manifestation of dynamics or chemistry? Nature, 322:811.
World Meteorological Organization/NASA, 1985, Assessment of our Understanding of the Processes Controlling its Present Distribution and Change.

THE ROLE OF ATOMIC AND MOLECULAR PROCESSES IN THE CRITICAL IONIZATION VELOCITY THEORY

Edmond Murad

Space Physics Division, Air Force Geophysics Laboratory

Hanscom AFB, MA, 01731, USA

INTRODUCTION

In formulating a theory for the structure of the solar system (meaning the formation of the planets with their individual satellites) Alfvén (1954; Alfvén and Arrhenius, 1975) postulated a simple intuitive concept, the Critical Ionization Velocity (CIV), to explain the condensation of matter in the early stages of the formation of the solar system. Since the original suggestion of CIV, it has been invoked to explain such diverse phenomena as cometary plasma (Formisano et al., 1982; Galeev et al.,1986) and the shuttle glow (Papadopoulos, 1984). Evidence, particularly from space experiments, has been reviewed recently by Newell (1985). Other reviews (e.g. Sherman, 1973) have emphasized the plasma (the collective) aspects of the theory. This review attempts to provide a transition from the collective aspects of the plasma treatments to the microscopic (collisional) implictions of the theory.

HISTORICAL BACKGROUND

In the early stages of the formation of the solar system the sun had largely formed; this proto-solar system still had a great deal of gas and condensed matter (in the form of grains and dust particles) throughout its medium. The question which intrigued Alfvén was how and why do the planets and their small satellites (e.g. the seventeen known moons of jupiter) form when the gravitational forces tend to attract the gaseous molecules towards the central body (either the sun or a given planet)?

As the atoms gain potential energy (because of the gravitational fall), their velocity will increase. Alfvén postulated that when the velocity reaches a value such that the resultant kinetic energy ($\frac{1}{2}mv^2$) equals the ionization potential ($e\Phi$), then the atom will be ionized, i.e.:

$$\tfrac{1}{2}mv^2 = e\Phi \qquad (1)$$

$$\text{or} \quad v = \sqrt{(2e\Phi/m)} \qquad (2)$$

where v is the velocity, $e\Phi$ the ionization potential, and m the mass. The ions and electrons thus produced will be trapped in the plasma because of the magnetic field, with the resulting effect that the density of the plasma will continue to increase. At the extreme, the density of the plasma is so high that the mean free path of the atoms will become smaller than the dimensions of the cloud, i.e. all infalling gas will be braked or captured by the plasma cloud. At that point, multiple collisions become frequent and condensation, leading eventually to the formation of the satellite, begins. In the the MKS system of units, Eq. (2) becomes $v = 13.9\sqrt{(IP/M)}$, where v is the velocity in km/s, IP the ionization potential in eV, and M the atomic mass in amu. Table 1 shows values of v for some of the elements. What is apparent from this table is that in terms of the critical ionization velocity, the elements fall into three groups: (I) hydrogen and helium; (II) the light elements (atomic number 3-10); and (3) heavy elements (atomic number > 10). This, in brief, is a summary of the origin of CIV.

Table 1. Critical Velocity Values for Some of the Elements

Element	Atomic Mass AMU	Ionization Potential (eV)	Critical Velocity km/s)
H	1	13.58	51.2
He	4	24.58	34.4
Li	6.9	5.39	12.2
Be	9	9.32	14.1
B	10.8	8.29	12.2
C	12	11.26	13.5
N	14	14.53	14.2
O	16	13.61	12.8
F	19	17.42	13.3
Ne	20.2	21.56	14.4
Na	23	5.14	6.6
Mg	24.3	7.65	7.8
Al	27	5.98	6.5
Si	28	8.15	7.5
P	31	10.48	8.1
S	32.1	10.36	7.9
Cl	35.5	12.97	8.4
Ar	39.9	15.76	8.7
K	39.1	4.34	4.6
Ca	40.1	6.11	5.4
Sc	44.9	6.54	5.3
Ti	47.9	6.82	5.2
V	50.9	6.74	5.1
Sr	87.6	5.70	3.5
Ag	107.9	7.58	3.7
Xe	131.3	12.13	4.2
Cs	132.9	3.89	2.4
Ba	137.3	5.21	2.7
La	138.9	5.58	2.8

The theory implies that ion-neutral reactions will initiate and propagate the condensation of the elements. How the initial ionization starts is not stated in the original formulation of the theory. The CIV theory seems to describe qualitatively the cosmology of the solar system.

EVIDENCE FOR THE CIV THEORY

Two types of experiments have been performed in order to test the validity of the CIV Theory: laboratory studies and space experiments. The results from these studies will be summarized separately below. It should be mentioned, however, that these experiments are quantitatively different from each other, as well as from the solar system (the object of the theory). The differences have been summarized by Möbius et al. (1979): (1) the laboratory studies have confining walls and the dimensions of the apparatus are sometimes smaller than the mean free path; (2) the densities in the laboratory experiments are much higher than those of the space experiments, and both are higher than the conditions for which the theory was proposed; and (3) the plasma is shot into the neutral gas rather than the reverse. Thus it remains an open question as to how to scale the laboratory experiments to the space experiments and how to scale the latter to the cosmic dimensions of the theory. Table 2 shows the plasma parameters for the three cases under discussion: laboratory experiments, space experiments, and cosmic conditions. Along with the experimental studies which will be described in this section, theoretical studies and modeling of the phenomena using computer simulation techniques have been undertaken. The two approaches have had a synergistic result, which will be described below.

Table 2. Comparison of CIV Parameters

	Solar System	Space Experiments	Homopolar Devices
B(tesla)	~ 1(-8)	3(-5)	0.01-1
[e] (cm^{-3})	10-1(3)	1(3)-1(5)	1(12)-1(15)
[n] (cm^{-3})	10-1(6)	1(5)-1(9)	1(13)-1(16)
r_e (m)	2(4)	~0.2	1(-5)
r_i (m)	2(7)	1(2)	1(-3)
λ_e (m)	2(8)	30	0.01
λ_i (m)	6(7)	100	3(-3)

[n,e] = electron and ion density
$r_{e,i}$ = gyroradii of electrons and ions under typical conditions
$\lambda_{e,i}$ = mean free path of electrons and ions under typical conditions

A. Laboratory Studies: The first experimental evidence for the theory came accidentally because of attempts to develop fusion machines. Alfvén (1960), Fahleson (1961), Baker et al. (1961), and Baker and Hammel (1961) have reported that voltage stored across a rotating plasma device (the homopolar device) is always limited to a certain critical value. Figure 1 shows a sketch of a homopolar device. Results from these devices indicate that following a short stagnation period (about 3 μs) the voltage reaches a value which lasts a long time (50-70 μs). This voltage is characteristic of the gas and is close to the velocity given by Eq. (1). Fahleson (1961) observed that the burning voltage is independent of gas pressure, depends on the gas (less for nitrogen than for hydrogen), is constant with increasing current and is approximately proportional to the magnetic field. At higher currents, the behavior becomes less regular. These experiments have been repeated numerous times, and the results have been quite reproducible, as summarized by the Danielsson (1973). Another laboratory CIV experiment has been used to study the interaction between a plasma accelerated from a conical theta pinch gun and a gas cloud (Danielsson, 1970). In this study radiation from He^* and He^+ were observed. Moreover, electrons with energy as great as 85 eV were observed. Danielsson (1970) and Danielsson and Brenning (1975) have extended the observations on CIV by using coaxial guns. In this work radiation from He* (501.5, 492.2, 438.8, and 471.3 nm) and He^+ (468.6 nm) was measured. Danielsson and Brenning (1975) shot a beam of Ar^+ through the plasma and observed the production of Ar^{++} using a mass spectrometer. These authors also measured the heating of electrons during the interaction of the plasma with the neutral gas; initially the electrons had energies of 5-10 eV, but following the interaction, they attained energies as high as 100 eV. The results also indicate that the plasma is stopped (due to the formation of ions via CIV) in less than 1 μs.

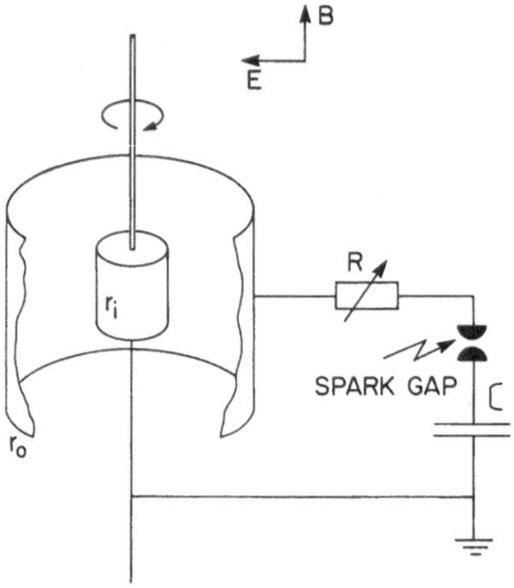

Figure 1 - Schematic drawing of a Homopolar device

Most of these experiments had heretofore concentrated on studying the CIV phenomenon in hydrogen or in rare gases. Axnäs (1978a) extended the work to other diatomic gases as well as to mixtures of gases. His experiments were carried out in a coaxial plasma gun. Molecular gases showed approximately the velocity expected from equation (1), although sometimes they were higher. Moreover, mixtures of gases seemed to have values of CIV which fall in between the values of the constituents, the exact value depending on the proportion of the gases in the mixtures.

A similar set of experiments was carried out by Mattoo and Venkataramani (1980) who studied the behavior of an argon plasma colliding with a hydrogen gas cloud, as well as a hydrogen plasma colliding with an argon gas cloud. They varied the kinetic energy of the plasma and observed that the plasma is stopped once the CIV of the neutral gas is reached. In a later study, Venkataramani and Mattoo (1980) observed that the plasma stream retards much more rapidly than can be explained by the CIV theory (assuming that ionization occurs by electron impact and assuming that the electron energy is 100 eV, as earlier studies had shown).

The interpretation of the laboratory experiments is not simple. Simpson (1981) analyzed the results from the homopolar devices and obtained the level of ionization observed without having to resort to anomalous processes. Aside from the interpretation, however, scaling these laboratory studies to space experiments or solar cosmology is not straightforward, as pointed out above and by other authors (Deehr et al., 1982, Möbius et al., 1979). The laboratory results are, in fact, somewhat perplexing in that they show the process to be quite efficient; indeed, much more efficient than can be explained by any classical collisional process (Danielsson, 1970).

B. Space Experiments: Newell (1985) has reviewed the experiments which have been conducted in space to verify CIV. The strongest evidence for the occurrence of this phenomenon in space comes from the PORCUPINE experiment by Haerendel (1982). Here Ba atoms were released in a grenade explosion, and a large fraction of these atoms had a velocity greater than the required CIV (2.7 km/s). Ba was released below the terminator, while the Ba^+ ions were observed by scattering of solar radiation above the terminator. In order for the ions to observed, they had to travel along the magnetic field lines; they were observed within 9 s of the release, much faster than the characteristic time for photoionization even in sunlit conditions (20 s) or charge exchange with O^+ (1000 s). Haerendel (1982) reported that nearly 30% of the barium atoms which had velocity greater than 2.7 km/s were ionized.

Deehr, et al., (1982) have reported the results of two experiments: one in which Ba was released in a shaped-charge explosion (KING CRAB), and another in which Ba doped with 1% Sr were released in a shaped explosion (BUBBLE MACHINE); Ba can be ionized either by CIV or by photoionization, whereas Sr cannot be photoionized fast enough. In both of these experiments conditions symptomatic of CIV, including ionization were observed. In BUBBLE MACHINE, Deehr et al. (1982) estimated that 16-50% of the Sr was ionized.

More recently, results of other space experiments have been published: The STAR of LIMA and the STAR of CONDOR. The STAR of LIMA (Wescott et al., 1986a; 1986b) consisted of a mother-daughter payload, where the daughter contained the Ba (CIV = 2.7 km/s) and the mother contained a full complement of plasma diagnostics. The STAR of CONDOR was similar, except that the explosive charge was Sr (CIV = 3.5 km/s). In addition to the plasma diagnostics on the mother rocket (e-field and magnetometer) optical observations from the ground and radar tracking were employed. The metal vapors were exploded in such a way that they travelled perpendicular to the earth's magnetic field lines. In the STAR of LIMA experiment some ionization was observed, and measurements of the electron density indicated that there was some heating of the electrons. However, the degree of ionization was considerably smaller (< 0.1%) than reported for PORCUPINE (Haerendel, 1982) and much less than expected from the laboratory experiments. In the STAR of CONDOR, the only expected source of ionization was CIV (photoionization and charge exchange being too slow); here no evidence was found for strong ionization. The 0.0036% ionization observed several minutes after the release could be explained by some thermal ionization during the release of Sr. The electric field measurements and the electron energy distributions (Kelley et al., 1986; Torbert and Newell, 1986) indicate that many of the prerequisites and many of the expected effects of CIV are present, although a discharge was not observed. The different space experiments thus yield a somewhat confusing set of data.

The arrival of the space shuttle and the availability of its bay for fairly sophisticated experiments at orbital velocity (7.3 km/s), other attempts have been made. Sasaki et al. (1986) have injected a fast (23 km/s) Ar plasma (consisting of a charge-neutral beam of Ar^+ and e) into the atmosphere surrounding the space shuttle. They reported observing radiation at 391.4 nm which they attributed to emission from the B state of N_2^+. They also observed VLF wave emissions near the lower hybrid frequencies of O^+, N_2^+, and Ar^+. Their gun was fired 253 times in 13 different flight configurations. In 7 of the 13 sequences no discharge was initiated. Moreover, when only neutral gas was injected into the atmosphere, ionization was not observed. Table 3 shows a summary of the space tests.

Table 3. Space Tests of the CIV Theory

Experiment	Location	Material	Percentage Ionization*	Reference
PORCUPINE	Kiruna	Ba	30%	Haerendel, 1982
KING CRAB	Alaska	Ba + <0.01% Sr	?	Deehr et al.,1982
BUBBLE MACHINE	Alaska	Ba + 1% Sr	16%?	Deehr et al.,1982
STAR OF LIMA	Peru	Ba	0.01%	Wescott et el.,1986b
STAR OF CONDOR	Peru	Ba + Sr	0.0036%	Wescott et al.,1986b
SEPAC	Shuttle	[Ar^++e] Plasma	See Text	Sasaki et al., 1986

* % of atoms having velocity greater than CIV

Other tests using the space shuttle have been proposed. Möbius et al. (1979) and Axnäs (1980) have proposed the release of Xe from the space shuttle; because of its high mass (131.3) and low ionization potential (12.1 eV), Xe has a relatively low CIV (4.2 km/s). More recently, Murad et al. (1986) have proposed the release of van der Waals molecules (e.g. dimers and trimers of N_2), which also have low CIV because of increased mass and lowered ionization potentials. Undoubtedly, some of these experiments will be performed.

MODEL

While the initial presentation of the theory (Alfvén, 1954) had a very intuitive approach to the problem, it provided a qualitative answer as to how planets and their moons developed in the initial stages of the formation of the solar system. Proceeding from a qualitative concept of CIV to a quantitative and rigorous theory have not been wholly successful, even though a number of attempts have been made; many improvements (e.g. Sherman, 1973; Raadu, 1978; Abe, 1984) have been made, however, since the early treatments. The best insight into the theory has come from computer simulations (see, for example, McBride et al., 1972,; Tanaka and Papadopoulos, 1983; Machida et al., 1984; Abe and Machida, 1985; Machida and Goertz, 1986).

These computer simulations have emphasized the plasma aspects of the theory, as opposed to the microscopic processes which play intermediate roles in the statistical aspects of the process. The most recent study, that of Machida and Goertz (1986), included a number of collisional processes in the computer simulation: a) elastic collisions between electrons and neutrals; b) excitation of allowed states by electron impact; c) ionizing collisions between electrons and neutrals; d) elastic collisions between ions and neutrals; e) charge exchange collisions between ions and neutrals; and f) photoionization. This computer simulation shows that electrons are heated by inelastic collisions with neutrals (ohmic heating) at high neutral densities and by trapping in lower hybrid waves (turbulent heating) at low neutral densities. The conclusion which Machida and Goertz reached was that anomalous ionization does occur when the relative velocity between the neutrals and the plasma exceeds the critical velocity.

The picture which emerges indicates that as the heated electrons collisionally ionize the neutrals, the different gyro-radii of the ions and electrons lead to a charge separation. A two-beam instability is established and leads to a plasma-type discharge. Figure 2 shows an idealized picture of a feedback loop for the CIV process. At every stage of the process there are loss terms which must be considered. Newell and Torbert (1986) have suggested that line excitation is an important loss term, which may explain why CIV was not observed in the STAR of LIMA and STAR of CONDOR experiments. Electron-ion recombination may also play a part in dissipating the energy of the electrons (Murad and Lai, 1986). While these loss terms will lead to optical radiation (a sign that CIV is taking place), they may also lead to quenching of the nascent discharge. Electron heating times (approximately 30 msec) derived from lower hybrid calculations (Tanaka and Papadopoulos, 1983) may thus be too optimistic.

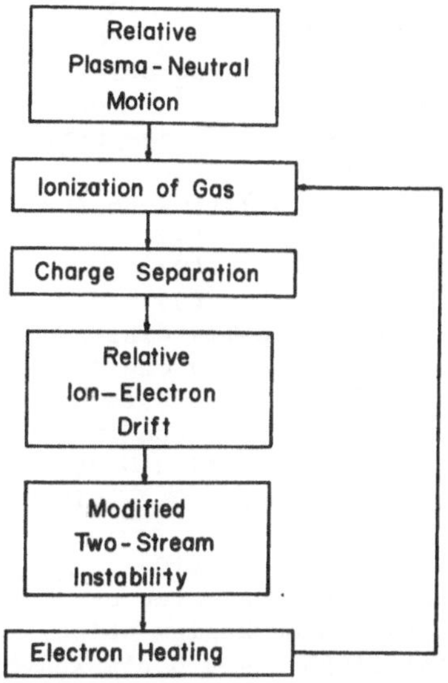

Figure 2 - Conceptual description of the CIV process

ATOMIC AND MOLECULAR PROCESSES IN CIV

This section raises more questions than it answers. Most of the questions have to do with understanding the nature of the experimental evidence for CIV, particularly the laboratory experiments. This is necessarily so because of the contradictory (or, at least, inconsistent) space evidence to date.

CIV is initiated when electrons in the magnetized plasma begin to gain kinetic energy. In the formation of the solar system, the ions are atomic, a fact which is conducive for the conversion of ion kinetic energy into electron kinetic energy since the recombination process:

$$A^+ + e \longrightarrow A + h\nu \qquad (3)$$

is rather improbable.

We will consider here only those processes which involve collisions. Electron heating by plasma mechanisms will not be discussed as more appropriate references are available (Sherman, 1973; Raadu, 1978; Tanaka and Papadopoulos, 1983; Machida and Goertz, 1986). Once the electrons are heated, a number of collisional processes begin to take place, as will be discussed below.

A. Electron-Neutral Collisions: The final process or processes leading to ionization (and, hence, to the initiation of a CIV discharge) are:

$$A + e \longrightarrow A^+ + 2e \qquad (4)$$

$$AB + e \longrightarrow AB^+ + 2e \qquad (5)$$

$$AB + e \longrightarrow A^+ + B + 2e \qquad (6)$$

for atomic and diatomic neutrals. Usually, the total electron impact ionization cross section, σ_4 or $\sigma_5 + \sigma_6$, has a maximum at an electron energy of ~ 100 eV (Kieffer and Dunn, 1966). At the maximum the cross section is ~ 10^{-15} cm^2. Electrons having energies of about 100 eV have been observed in laboratory experiments on CIV (Danielsson and Brenning, 1975). The rate constant for ionization is ~ 10^{-9} cm^3 s^{-1}. For the data given in Table 2, this rate constant would lead to a characteristic ionization time, τ, for the neutrals of 10^8-10^3, 10^4-1, and 10^{-4}-10^{-7} s for the solar system, space experiments, and laboratory experiments, respectively.

In addition to (4)-(6), other electron-neutral reactions can take place:

$$e + A \longrightarrow A^* + e \qquad (7)$$

$$e + A \longrightarrow A^{**} + e \qquad (8)$$

$$e + AB \longrightarrow A + B + e \qquad (9)$$

where A* is a long-lived, electronically excited (metastable) state of A, and A** is a short-lived, electronically excited state of A. Reactions similar to (7) and (8) would be applicable to diatomic molecules. In any case, A* would tend to propagate CIV, since its long lifetime can lead to energy pooling and ionization (see section C below). On the other hand, both (8) and (9) would retard or quench CIV, since in one case the energy is dissipated by radiation:

$$A^{**} \longrightarrow A + h\nu \qquad (10)$$

and in the other, (9), energy is dissipated by dissociation of AB. Indeed, (8) + (9) is the line excitation mechanism which, according to the calculations of Newell and Torbert (1985), would dissipate 30% of the energy.

For a measure of the magnitudes of the processes mentioned in this section, we consider the case where the neutral is nitrogen, N_2. σ_7 where N_2^* is the A $^3\Sigma$, is ~ 0.25 x 10^{-16} cm^2 at an electron energy of about 17 eV (Trajmar and Cartwright, 1984). σ_8 where N_2 is the c' $^1\Sigma$ state, is ~ 0.15 x 10^{-15} cm^2 (Trajmar and Cartwright, 1984). By comparison, σ_9 is ~ 1 x 10^{-16} cm^2.

B. Electron-Ion Collisions: These reactions, such as (3) and the dissociative recombination reaction for diatomic molecular ions:

$$AB + e \longrightarrow A + B + h\nu \qquad (11)$$

can play a role in quenching CIV discharge. Reaction (11) may also be a source of radiation from excited fragments (Murad and Lai, 1986). The important point to keep in mind is that k_3 is ~ 10^{-12} cm^3 $molec^{-1}$ s^{-1} (Bates and Dalgarno, 1962; Dunn et al., 1984), while k_{11} is ~ 10^{-6}-10^{-7} cm^3 $molec^{-1}$ s^{-1} (Bates, 1979; McGowan and Mitchell, 1984). The characteristic time for neutralization by reaction (3), τ_3, is 10^{11}-10^9, 10^9-10^7, and 1-10^{-3} s, respectively for solar system, space experiments, and homopolar devices. By contrast, for diatomic (or polyatomic) molecules τ_{11} is 10^6-10^4, 10^4-10^2, and 10^{-5}-10^{-8} s, respectively for solar system, space experiments, and homopolar devices. Thus for laboratory experiments involving molecular species reaction (12) may well be important in terminating the CIV process. At the electron densities reported for laboratory experiments (see Table 2), three-body recombination can take place:

$$A^+ + e + e \longrightarrow A^* + e \qquad (3a)$$

At high neutral densities (e.g. [n] ~ 10^{13} cm^{-3}) k_{3a}, the rate constant for the three body recombination is greater than k_3 at electron temperatures as high as 4000 K (Bates et al., 1962a; Bates et al., 1962b; Bates, 1979). Reaction (3a) would be expected to quench high density discharges as well as to generate radiation from excited neutral products.

C. Neutral-Neutral Collisions: An important process to consider is that of energy pooling through the formation of metastable electronic states, A* in reaction (7) above. A* may collide with another A*, because of the high densities to yield a species which is easily ionized, viz:

$$A^* + A^* \longrightarrow A + A^\dagger \qquad (12)$$

where $A^\dagger$ is a highly excited state, such that it can be easily ionized by collisons with low energy electrons:

$$A^\dagger + e \longrightarrow A^+ + e \qquad (13).$$

Lai et al. (Work in progress) have, in fact, considered processes such as the sequence of (7), (12) and (13) as a way of obtaining faster ionization. Their results indicate that for both He and N_2, such processes can accelerate the ionization a great deal. What helps is that the lifetime of the metastable states are quite long (He 2 ^3S at 7900 s and N_2 $A^3\Sigma^+$ at 1.3 s). Deactivation by collisions with electrons was included in the calculations mentioned above:

$$A^* + e \longrightarrow A + e_{fast} \qquad (14)$$

D. Ion-Neutral Reactions: Ion-neutral collisions should also play a part in the propagation and condensation (the latter leading, perhaps, to quenching of the discharge); this is particulary so when molecular gases are involved (Axnäs, 1978). For example, in the case of N_2 the following reactions should also be considered:

$$N_2^+ + N_2 \longrightarrow N_3^+ + N \qquad (15)$$

$$N_2^+ + 2\,N_2 \longrightarrow N_4^+ + N_2 \qquad (16)$$

σ_{15} and σ_{16} have been measured as a function of energy by a number of authors (e.g. Maier,II, 1971; 1974; Bowers, et al., 1974; Rowe et al., 1984; van Koppen et al., 1984). These measurements may have been made with excited states in reaction (15). σ_{15} has a maximum at E_{CM} ~ 12 eV, the maximum cross section being about 5×10^{-18} cm^{-2} (Maier, II, 1974), which translates, roughly, into a rate constant of ~ 5×10^{-12} cm^3 $molec^{-1}$ s^{-1}. At an ion (or electron) density of ~ 1×10^{13} cm^{-3}, τ_{15} ~ 0.02 s. Reaction (16) is quite fast. Even at at T = 1000 K, k_{12} is ~ 2(-29) cm^6 $molecule^{-2}$ s^{-1} (van Koppen et al., 1984). Reactions of this type clearly have to be included in order to be able to understand the laboratory experiments on CIV, which, in turn is needed in order to have a rigorous definition of the CIV theory.

SUMMARY

This review has presented a summary of a theory for which there is considerable experimental evidence. Most of the earlier theoretical treatments had attempted to explain the theory in terms of statistical (plasma) concepts. Our aim here was to emphasize the role of microscopic (collisional) processes in the understanding of the theory. The experiments which seem to confirm the theory pose a number of questions about the interpretation. For example, the laboratory experiments show that there are large concentrations of ions and electrons, seemingly without their neutralizing each other, even though the dimensions of the apparatus are larger than the mean free paths. In addition to these experiments it is important to understand the ionization mechanism in mixtures of gases. The findings by Axnäs (1978a) that values of the critical velocity for mixtures fall between those of the components are consistent; yet there is no simple explanation for this, if the ionization process is electron impact ionization. Computer simulations will have to be repeated with the inclusion of processes such as (3), (3a) and (7)-(16) before a realistic measure of the agreement of theory and experiment can be obtained.

Finally, it is crucial to establish the conditions under which this theory works in space. As of this writing there has been only one experiment, PORCUPINE, which has yielded positive results. It is important to establish why this experiment gave such unique results. This task becomes crucial for a number of applications, particularly as we move towards permanently-manned space stations. Whether, and under what conditions, are manned stations in low earth orbit surrounded by plasma needs to be established.

ACKNOWLEDGEMENTS

I would like to thank Drs. Ingvar Axnäs, William F. Denig, Irving Kofsky and Shu Lai for helpful discussions.

REFERENCES

Abe, T., 1984, Theory for the Critical Ionization Velocity Phenomenon, Planet. Space Sci., 32:903.

Abe, T., and Machida, S., 1985, Production of High Energy Electrons Caused by Counterstreaming Ion Beams in an External Magnetic Field, Phys. Fluids, 28:1178.

Alfvén. H., 1954, The Origin of the Solar System, Oxford U. P., Oxford.

Alfvén, H., and Arrhenius, G., 1975, Structure and Evolutionary History of the Solar System, Reidel, Dordrecht.

Alfvén, H., 1960, Collision Between a Nonionized Gas and a Magnetized Plasma, Revs. Mod. Phys., 32:710.

Axnäs, I., 1972, Experimental Investigation of an Ionizing Wave in a Coaxial Plasma Gun, TRITA-EPP-72-31, Report from the Royal Institute of Technology, Stockhom, Sweden.

Axnäs,I., 1978a, Experimental Investigation on the Critical Ionization Velocity in Gas Mixtures, Astrophys. Space Sci., 55:139.

Axnäs, I., 1978b, Experimental Comparison of the Critical Ionization Velocity in Atomic and Molecular Gases, Report from Royal Institute of Technology, Tritta-EPP-78-04.

Axnäs, I., 1980, Some Necessary Conditions for a Critical Velocity Interaction Between the Ionospheric Plasma and a Xenon Cloud, Geophys. Res. Lett., 7:933.

Baker, D. A., Hammel, J.E., and Ribe, F., 1961, Rotating Plasma Experiments. I. Hydromagnetics Properties, Phys. Fluids, 4:1534.

Baker, D. A., and Hammel, J. E., 1961, Rotating Plasma Experiments. II. Energy Measurements and the Velocity Limiting Effect, Phys. Fluids, 4:1549.

Bates, D. R., 1979, Aspects of Recombination, in Advances in Atomic and Molecular Physics, Vol. 15, Edited by D. R. Bates, Academic Press, NY, 235-261.

Bates, D. R., and Dalgarno, A., 1962, Electronic Recombination, in Atomic and Molecular Processes, Edited by D. R. Bates, Academic Press, NY, 245.

Bates, D. R., Kingston, A. E., and McWhirter, R. W. P. ,1962a, Recombination Between Electrons and Atomic Ions. I. Optically Thin Plasmas, Proc. Roy. Soc. (London), A267:297.

Bates, D. R., Kingston, A. E., and McWhirter, R. W. P., 1962b, Recombination Between Electrons and Atomic Ions. II. Optically Thick Plasmas, Proc. Roy. Soc. (London), A270:155.
Bowers, M. T., Kemper, P. R., and Laudenslager, 1974, Reactions of Ions in Excited Electronic States: $(N2^{+}.)^{*} + N_2 \text{--->} N_3^{+} + N$, J. Chem. Phys., 61:4394.
Brenning, N., 1980, Electron Temperature Measurements in Low-Density Plasmas by Helium Spectroscopy, J. Quant. Spectrosc. Rad. Transfer, 24:293.
Danielsson, L., 1970, Experiment on the Interaction Between a Plasma and a Neutral Gas, Phys. Fluids, 13:2288.
Danielsson, L., 1973, Review of the Critical Velocity on Gas Plasma Interaction.I.Experimental Observations, Astrophys. Space Sci., 24:459.
Danielsson, L., and Brenning, N., 1975, Experiment on the Interaction Between a Plasma and a Neutral Gas. II., Phys. Fluids, 18:661.
Deehr, C. S.,Wescott, E. M., Stenback-Nielsen, H., Romick, G. J., Hallinan, T. J., and Föppl, H., 1982, A Critical Velocity Interaction Between Fast Barium and Strontium Atoms and the Terrestrial Ionospheric Plasma, Geophys. Res. Lett., 9:195.
Dunn, G. H., Belic, D. S., Morgan, T. J., Mueller, D. W., and Timmer, C., 1984, Dielectronic Recombination of Some Single-Charge Ions, in Electronic and Atomic Collisions, Edited by J. Eichler, I. V. Hertel, and N. Stollerhoft, Elsevier, Amsterdam, 809-817.
Fahleson, U. V., 1961, Experiments with Plasma Moving Through Neutral Gas, Phys. Fluids, 4:123.
Formisano, V., Galeev, A. A., and Sagdeev, R. Z., 1982, The Role of the Critical Ionization Velocity Phenomena in the Production of Inner Coma Cometary Plasma, Planet. Space Sci., 30:491.
Galeev, A. A., Gringauz, K. I., Klimov, S. I., Remizov, A. P., Sagdeev, R. Z., Savin, S. P., Sokolov, A. Yu., and Verigin, M. I., 1986, Critical Ionization Velocity Effects in the Inner Coma of Comet Halley: Measurements by Vega-2, Geophys. Res. Lett., 13:845.
Haerendel, G., 1982, Alfvén's Critical Velocity Effect Tested in Space, Z. Naturforschung, A37:728.
Kieffer, L. J., and Dunn, G. H., 1966, Electron Impact Ionization Cross Section Data: Atoms, Atomic Ions, and Diatomic Molecules. I. Experimental Data, Revs. Mod. Phys., 38:1.
Kelley, M.C., Pfaff, R. F., and Haerendel, G., 1986, Electric Field Measurements During the Condor Critical Velocity Experiment, J. Geophys.Res., A91:9939.
Lai, S., McNeil, W. J., and Murad, E., Work in Progress.
Machida, S., Abe, T., and Terasawa, T., 1985, Computer Simulation of Critical Velocity Ionization, Phys. Fluids, 27:1928.
Machida, S., and Goertz, C. K., 1986, A Simulation Study of the Critical Ionization Velocity Process, J. Geophys. Res., A91:11965.
Maier, II, W. B., 1971, Reactions Between N_2^{+} and N_2, J. Chem. Phys., 55: 2699.
Maier, II, W. B., 1974, Reactions Between Isotopically Labeled N_2^{+} and N_2 for Primary Ion Energies Below 45 eV, J. Chem. Phys., 61:3459.
Mattoo, S. K., and Venkataramani, N., 1980, On the Threshold Velocity in the Interaction Between a Magnetized Plasma and a Neutral Gas, Phys.Lett., 76A:257.
McBride, J. B., Ott, E., Boris, J. P., and Orens, J. H., 1972, Theory and Simulation of Turbulent Heating by Modified Two-Stream Instability, Phys. Fluids, 15:2367.
McGowan, J. W., and Mitchell, J. B. A., Electron Molecular Positive Ion Recombination, in Electron-Molecule Interactions and Their Applications, Vol. 2, Edited by L.G. Christophorou, Academic Press, Orlando, FL, 65.
Möbius, E. Boswell, R. W., Piel, A., and Henry, D., 1979, A Spacelab Experiment on the Critical Ionization Velocity, Geophys. Res. Lett., 6:29.

Murad, E., Lai, S. T., and Stair, Jr., A. T., 1986, A Proposed Experiment to Study the Critical Ionization Velocity Theory in Space, J. Geophys. Res., A91:10188.

Murad, E., and Lai, S., 1986, Effect of Dissociative Electron-Ion Recombination on the Propagation of Critical Ionization Dishcarges, J. Geophys. Res., A91:13745.

Newell, P. T., 1985, Review of the Critical Ionization Velocity Effect in Space, Revs. Geophys., 23:93.

Newell, P. T., and Torbert, R. B., 1985, Competing Processes in Ba and Sr Injection Critical Ionization Velocity Experiments, Geophys. Res. Lett., 12:835.

Papadopoulos, K., 1984, On the Shuttle Glow (the Plasma Alternative), Radio Science, 19:571.

Rowe, B. R., Dupeyrat, G., Marquette, J. B., and Gaucherel, P., 1984, Study of the Reactions $N_2^+ + 2 N_2 \longrightarrow N_4^+ + N_2$ and $O_2^+ + 2 O_2 \longrightarrow O_4^+ + O_2$ from 20 to 160 K by the CRESU Technique, J. Chem. Phys., 80:4915.

Sasaki, S., Kawashima, N., Kuriki, K., Yanagisawa, M., Obayashi, T., Roberts, W. T., Reasoner, D. L., Taylor, W. W. L., Williams, P. R., Banks, P. M., and Burch, J. L., 1986, Gas Ionization Induced by a High Speed Plasma Injection in Space, Geophys. Res. Lett., 13:434.

Sherman, J. C., 1973, Review of the Critical Velocity of Gas-Plasma Interaction, Astrophys. Space Sci., 24:487.

Simpson, S. W., 1981, A Steady State Fluid Model of a Rotating Plasma, Phys. Fluids, 24:418.

Tanaka, M., and Papadopoulos, K., 1983, Creation of High Energy Tails by Means of the Modified Two-Stream Instability, Phys. Fluids, 26: 1697.

Torbert, R. B., and Newell, P. T., 1986, A Magnetospheric Critical Velocity Experiment: Particle Results, J. Geophys. Res., A91:9947.

Trajmar, S., and Cartwright, D. C., 1984, Excitation of Molecules by Electron Impact, in Electron-Molecule Interations and Their Applications, Vol. 1, Edited by L. G. Christophorou, Academic, Orlando, FL. 155.

van Koppen, P. A. M., Jarrold, M. F., Bowers, M. T., Mass, L. M., and Jennings, K. R., 1984, Ion-Molecule Association Reactions: A Study of the Temperature Dependence of the Reaction $N_2^{+\cdot} + N_2 + M \longrightarrow N_4^{+\cdot} + M$ for $M = N_2$ and He: Experiment and Theory, J. Chem. Phys., 81:288.

Venkataramani, N., and Mattoo, S. K., 1980, Plasma Retardation in Alfvén's Critical Velocity Phenomenon, Phys. Lett., 79A:393.

Wescott, E. M., Stenbaek-Nielsen, H.C., Hallinan, T., Föppl, H. and Valenzuela, A., 1986a, Star of Lima: Overview and Optical Diagnostics of a Barium Alfvén Critical Velocity Experiment, J. Geophys. Res., A91:9923.

Wescott, E. M., Stenbaek-Nielsen, H.C., Hallinan, T., Föppl, H. and Valenzuela, A., 1986b, Star of Condor: A Strontium Critical Velocity Experiment, Peru, 1983, J. Geophys. Res., A91:9933.

Zipf, E. C., 1984, Dissociation of Molecules by Electron Impact, in Electron-Molecule Interactions and Their Applications, Vol. 1, Edited by L. G. Christophorou, Academic, Orlando, FL. 335.

SPECTRAL SOLAR IRRADIANCES AND AERONOMIC PHOTOLYTIC PROCESSES IN THE MESOSPHERE

Marcel Nicolet

Space Aeronomy Institute
3 Avenue Circulaire
B-1180 Brussels, Belgium

INTRODUCTION

The advances in space observations during the past thirty years have overshadowed many fundamental aspects of aeronomy which dominated the first days of rocket observations. In particular, the mesosphere has suffered from a lack of attention before the development of the MAP (Middle Atmosphere Programme). On the other hand, the mass of material obtained by the SME (Solar Mesosphere Explorer) satellite should lead the progress in our understanding of the mesosphere with its various aeronomic processes.

The principal gases in the thermosphere, molecular nitrogen, atomic oxygen and molecular oxygen, limit the atmospheric penetration of solar radiation at wavelengths less than 100 nm. Since absorption cross-sections are not less than 10^{-18} cm^2 between 100 nm and 10 nm, the solar radiation in this spectral range is absorbed above 100 km and ionizes N_2, O and O_2 and other minor constituents.

At 175 nm, the O_2 Schumann-Runge continuum begins and its absorption leads to the direct photodissociation of that molecule. Thus, the total number of photons available at the top of the earth's atmosphere at wavelengths less than 175 nm represents the total number of oxygen molecules which can be photodissociated in a vertical column of the atmosphere. The continuum of the Schumann-Runge system has an absorption cross-section varying from about 10^{-17} cm^2 at the absorption peak to more than 10^{-19} cm^2 at the threshold. Therefore, the penetration of solar radiation into the atmosphere is limited to the thermosphere and cannot reach the mesosphere. But, it must be realized (Nicolet and Mange, 1954; Nicolet, 1954; Mange, 1955, 1957, 1961) first that the dissociated atoms move downward and recombine below 100 km, above the mesopause, and second that molecular oxygen is not in photochemical equilibrium and is subject to an upward transport trying to reach diffusive equilibrium conditions. Furthermore, the effective recombination of the oxygen atoms is still an unsolved problem since the O concentration peak in the neighbourhood of 95 km (see, for example, Howlett et al., 1980; Dickinson et al., 1980; Sharp, 1980; Offerman et al., 1981; Dickinson et al., 1985) is an aeronomic parameter with a value varying even outside of the limits of 10^{11} and 10^{12} atoms cm^{-3}.

Finally, the solar irradiance and its variation with solar activity in the spectral range of the O_2 Schumann-Runge continuum is not yet an aeronomic parameter known with the required accuracy. Figure 1 gives the observational results (Nicolet, 1981) which were obtained between 1970 and 1980, after the analysis of the first observational results made by Ackerman (1971). However, if recent observations such as the SME solar irradiances (Rottman, to be published) or rocket observations (Mentall, to be published) are considered, it seems possible to conclude that the lowest observational values of the solar flux for quiet and active sun conditions may be accepted with a variation with solar activity smaller than a factor of 2.

In conclusion, the production of oxygen atoms in the mesosphere clearly includes not only the absorption by O_2 beginning at 240 nm in the Herzberg continuum, where the value of the cross section has always been acknowledged as uncertain due to the experimental pressure effects in the laboratory determination, but also the extremely variable absorption between 200 and 175 nm by the Schumann-Runge bands as a result of their predissociation. These aspects will be taken into consideration in our analysis with other aspects such as the penetration of solar radiation to greater depths than 10^{19} O_2 molecules cm^{-2} occurring in various "windows" between 122 and 110 nm. An important window with a cross section of the order of 10^{-20} cm^2 is situated at 121.6 nm corresponding to the solar-chromospheric Lyman-alpha line of atomic hydrogen and is responsible for various photodissociation processes in the mesosphere.

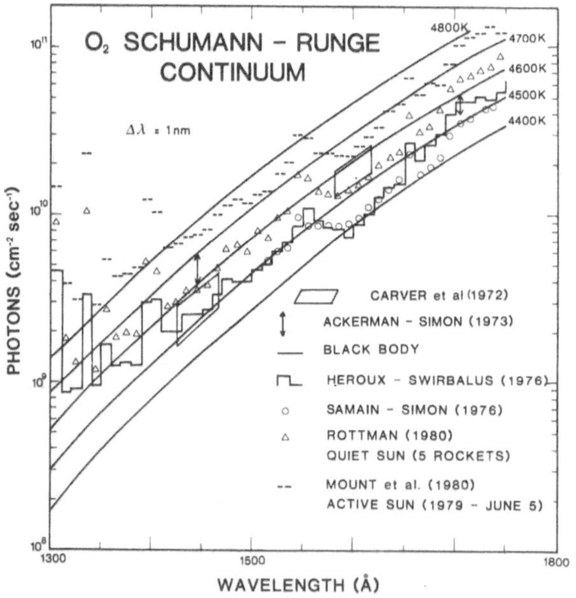

Figure 1

Spectral distribution of the solar irradiance in the region corresponding to the Schumann-Runge continuum of molecular oxygen leading for $\lambda < 175$ nm to about 5×10^{11} photons cm^{-2} s^{-1} (possible minimum value) to 1.5×10^{12} photons cm^{-2} s^{-1} (too high maximum value probably to be reduced to a possible maximum of about 1×10^{12} cm^{-2} s^{-1}). See references in Nicolet (1981).

THE CHEMICAL REACTIONS IN THE MESOSPHERE

When Chapman (1930a) introduced the atomic oxygen region, he had in mind a possible explanation for the airglow emission of the green line that arises from a forbidden transition of atomic oxygen. The photodissociation was due to the absorption of solar radiation (a black body at 6000K) in the O_2 Schumann-Runge continuum, a thermospheric process since the Herzberg continuum was not yet known:

$$(J_2);\quad O_2 + h\nu\ (\lambda < 175\ \text{nm}) \rightarrow 2O \qquad (1)$$

with recombination of oxygen atoms in the presence of a third body (N_2, O_2, O)

$$(k_1);\quad O + O + M \rightarrow O_2 + M \qquad (2)$$

and with the possibility (M=O) to excite the third body $O(^3P)$ in the metastable state $O(^1S)$.

In addition, Chapman (1930b) introduced also for his study of atmospheric ozone the association of oxygen atoms with oxygen molecules

$$(k_2);\quad O + O_2 + M \rightarrow O_3 + M \qquad (3)$$

with a possible photodissociation (Hartley band)

$$(J_3);\quad O_3 + h\nu(\lambda < 300\ \text{nm}) \quad O_2 + O \qquad (4)$$

and also the two-body reaction between O_3 and O

$$(k_3);\quad O + O_3 \rightarrow 2O_2 \qquad (5)$$

leading to the re-formation of oxygen molecules and the final destruction of ozone.

The discovery by Meinel (1950) that the vibrational rotational bands of the hydroxyl radical OH appear in the airglow aroused the interest in the photochemistry of hydrogen-oxygen compounds (Bates and Nicolet 1950a,b), and in particular, of water vapour (Bates and Nicolet, 1950c). Their results were based on very rough estimates of the various aeronomic parameters involved. Figure 2 gives their results for the mesospheric distributions of the various forms of hydrogen: atomic hydrogen, molecular hydrogen, the hydroxyl and per-hydroxyl radicals, water vapour. Their analysis not only explained the observations on the airglow but had several other points of interest such as the presence of hydrogen reducing the equilibrium concentrations on both O and O_3, the presence of hydrogen in atomic and molecular form (H and H_2) leading to the upward escape of hydrogen into interplanetary space and also the rapid decrease of water vapour concentration in the upper mesosphere by increasing photodissociation. But, it should be remembered that these results were only rough estimates based on photochemical equilibrium conditions with absorption cross sections determined by extrapolation in various parts of the O_2 spectrum (without introducing the predissociation in the O_2 Schumann-Runge bands) and, at that time with several unknown reaction rate coefficients.

The principal reactions in a hydrogen-oxygen mesosphere start from the dissociation of water vapour by various photodissociative channels (see Nicolet, 1984 and references infra)

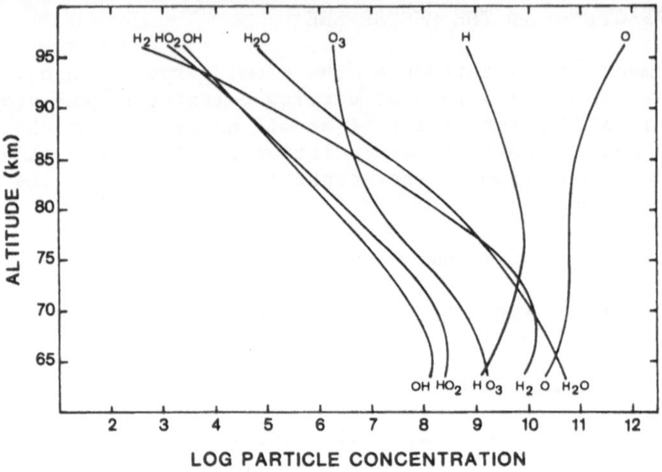

Figure 2

Approximate mesospheric distributions under photochemical equilibrium conditions of atomic oxygen and ozone in an oxygen-hydrogen atmosphere with its additional constituents H_2O, H_2, H, OH and HO_2. After Bates and Nicolet (1950); first aeronomic analysis of hydrogen compounds in the terrestrial atmosphere.

$$(J_{OH-H}) \;;\; H_2O + h\nu(\lambda\lambda 200\text{-}175\ \text{nm}) \rightarrow OH(X^2\Pi) + H(^2S) \tag{6}$$

$$(J_{OH-H}) \;;\; H_2O + h\nu(\text{Ly-}\alpha) \rightarrow OH(X^2\Pi) + H(^2S) \;;\; 70\% \tag{7}$$

$$(J_{OH^*-H}) \;;\; \rightarrow OH(A^2\Sigma) + H(^2S) \;;\; 8\% \tag{8}$$

$$(J_{H_2-O}) \;;\; \rightarrow H_2(X^1\Sigma_g^+) + O(^1D) \;;\; 10\% \tag{9}$$

$$(J_{H-H-O}) \;;\; \rightarrow O(^3P) + 2H(^2S) ;\; 12\% \tag{10}$$

With the production of H atoms and OH radicals, various reactions are possible (see Figure 3). In the mesosphere the principal processes are

$$(a_1) \;;\; H + O_2 + M \rightarrow HO_2 + M \tag{11}$$

the production of the perhydroxyl radical and

$$(a_2) \;;\; H + O_3 \rightarrow O_2 + OH^* \tag{12}$$

the production of the excited hydroxyl radical leading to the airglow emission in the infrared.

These two reactions are followed by processes involving OH and HO_2

$$(a_5) \;;\; O + OH \rightarrow H + O_2 \tag{13}$$

and

$$(a_7) \;;\; O + HO_2 \rightarrow OH + O_2 \tag{14}$$

i.e. the first introduction of a catalyst (hydrogen) leading to a loss process of atomic oxygen (and ozone).

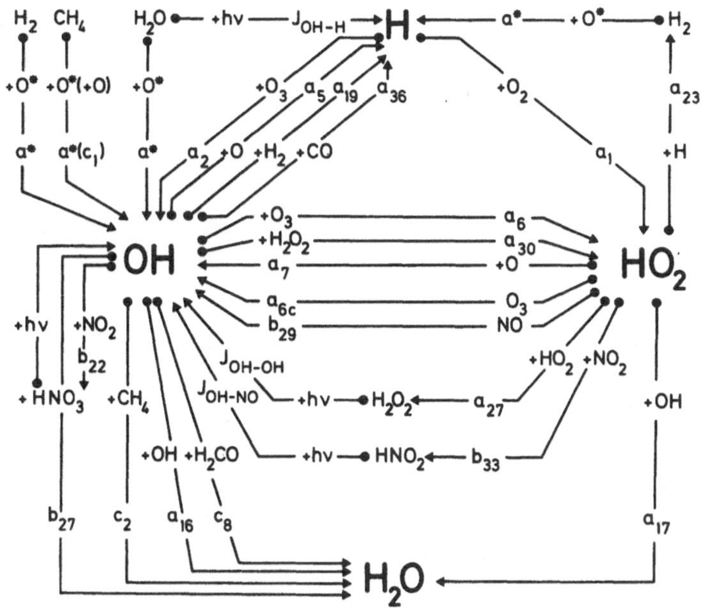

Figure 3

Fundamental reaction scheme in which atomic hydrogen (H), hydroxyl radical (OH) and perhydroxyl radicals (HO_2) may be involved. Their reactions with ozone and nitrogen oxides are essentially stratospheric processes. CH_4 and H_2O_2 do not play a significant role in the mesosphere.

After its photodissociation, H_2O can be re-formed by the reaction

$$(a_{17}) \; ; \; OH + HO_2 \rightarrow H_2O + O_2 \qquad (15)$$

Reactions with O_3 or leading to H_2O_2 can be neglected at mesospheric levels. Reactions between H and HO_2

$$(a_{15}) \; ; \; H + HO_2 \rightarrow OH + OH \qquad (16)$$

$$(a_{20}) \; ; \; \rightarrow H_2O + O \qquad (17)$$

$$(a_{23}) \; ; \; \rightarrow H_2 + O_2 \qquad (18)$$

cannot play an important role in the general behaviour of the mesosphere. Reaction (18) must be associated with the photodissociation process (9), both being a mesospheric source of molecular hydrogen.

THE CHEMICAL EQUATIONS IN THE MESOSPHERE

The equation governing the rate of change of the concentration of an atmospheric constituent XY is of the form

$$\frac{\partial n(XY)}{\partial t} + \text{div}\,[n(XY)w(XY)] = P - L \qquad (19)$$

where P and L denote the production and loss terms due to the photochemical or chemical processes, respectively, and div [nw] corresponds to the transport term which is always introduced when atmospheric exchanges are more significant than chemical processes. This transport term is often used now as a significant mechanism responsible for apparently inoperative aeronomic processes.

The rate of change of n(O) in the mesosphere should be written (see, for example, Nicolet, 1979) in general form

$$
\begin{aligned}
&\frac{\partial n(O)}{\partial t} + \mathrm{div}^{-}[n(O)w(O)] \\
&+ 2k_{1M}n^2(O) + k_{2M}n(O_3)n(O) \\
&+ n(O)\ [k_3n(O_3) + a_5n(OH) + a_7(HO_2) + a_{24}n(H_2)] \\
&+ n(O^*)\ [a^*_{H_2O}n(H_2O) + a^*_{H_2}\ n(H_2)] \\
&= 2n(O_2)J_2 + n(O_3)J_3 + n(H_2O)\ [J_{H_2-O} + J_{2H-O}] + n(HO_2)J_{HO_2} \\
&+ n(H)\ [a_{20}n(HO_2) + a_{22}n(OH)]
\end{aligned}
\tag{20}
$$

where O^* is the excited oxygen atom in the metastable state ^1D.

The associated equation to (20) corresponding to the rate of change of $n(O_3)$ in the mesosphere is

$$\frac{dn(O_3)}{dt} + n(O_3)\ [J_3 + k_3n(O) + a_2n(H)] = k_{2M}n(O_2)n(O) \tag{21}$$

In these equations the reactions of OH and HO_2 with O_3 have been neglected since they are compared with their reactions with O in the combination of the two equations (20) and (21).

The practice which is followed throughout this article is to write $\langle XY \rangle$ instead of $\frac{\partial n(XY)}{\partial t}$ + div [n(XY)w(XY)] or only div [n(XY)w(XY)] for a stationary state and {XY} instead of $\frac{dn(XY)}{dt}$ = P - L for the rate of change of the concentration n(XY). When {XY} = 0, there is photoequilibrium.

The practical equation for the mesosphere is, therefore, $\langle O \rangle + \{O_3\}$

$$
\begin{aligned}
&+ 2k_{1M}n^2(O) + n(O_3)\ [2k_3n(O) + a_2n(H)] \\
&+ n(O)\ [a_5n(OH) + a_7n(HO_2) + a_{24}n(H_2)] \\
&+ n(O^*)\ [a^*_{H_2O}n(H_2O) + a^*_{H_2}n(H_2)] \\
&= 2n(O_2)J_2 + n(H_2O)\ [J_{H_2-O} + J_{2H-O}] + n(HO_2)J_{HO_2} \\
&+ n(H)\ [a_{20}n(HO_2) + a_{22}n(OH)]
\end{aligned}
\tag{22}
$$

where $k_{1M} = k_1n(M)$ and $k_{2M} = k_2n(M)$.

The rate of change of the concentration of water vapour in the mesosphere can be written:

$$<H_2O> + n(H_2O)[J_{OH-H} + J_{2H-O} + J_{H_2-O} + a^*_{H_2O}n(\rho^*)]$$
$$= a_{20}n(H)n(HO_2) + a_{17}n(OH)n(HO_2) \quad (23)$$

The water vapour dissociation can be described in the mesosphere by its photodissociation and at the stratopause level by its reaction with the excited atom $O(^1D)$. H_2O is subject to a permanent transport in the mesosphere related to its photodissociation and reformation from the reaction between OH and HO_2.

The associated equation for the hydroxyl radical is

$$\{OH\} + n(OH)\ [a_5n(O) + a_{17}n(HO_2)]$$
$$= n(H)\ [a_2n(O_3) + 2a_{15}n(HO_2)] + n(HO_2)\ [a_7n(O) + J_{HO_2}]$$
$$+ n(H_2)a^*_{H_2}\ n(O^*) + n(H_2O)\ [J_{OH-H} + 2a^*_{H_2O}n(O^2)] \quad (24)$$

The last three equations (22), (23) and (24) lead to an expression determining the mesospheric conditions:

$$<O> + \{O_3\} = <H_2O> + \{OH\}$$
$$+ 2k_{1M}n^2(O) + 2k_3n(O)n(O_3) + 2a_5n(O)n(OH)$$
$$= 2n(O_2)J_2 + 2n(HO_2)\ [J_{HO_2} + \{a_{15} + a_{20}\}n(H)] \quad (25)$$

The terms involving HO_2 can be neglected compared with the atomic oxygen production by O_2 photodissociation. Furthermore, if the photo-equilibrium conditions (low solar zenith angles) are applied to O_3 and OH, i.e. $\{O_3\} = 0$ and $\{OH\} = 0$ in equation (25), then

$$a_7n(O)n(HO_2) = a_{1M}n(O_2)n(H) \quad (26)$$

and

$$\frac{n(OH)}{n(HO_2)} = \frac{a_7}{a_5}\ [1 + \frac{a_2n(O_3)}{a_{1M}n(O_2)}] \quad (27)$$

while

$$\frac{n(O_3)}{n(O)} = \frac{k_{2M}n(O_2)}{J_3 + a_2n(H) + k_3n(O)} \quad (28a)$$

In applying numerical values in (28a) for a standard atmosphere, it can be seen that the ratio $n(O_3)/n(O) = 1$ near 60 km (daytime conditions), i.e. in the lower mesosphere above the stratopause level where photo-equilibrium conditions can be 9550 med for almost all aeronomic parameters. Furthermore, $k_3n(O) < J_3+a_2n(H)$ and instead of (28a) we see that

$$n(O_3)/n(O) = k_{2M}n(O_2/J_3 + a_2n(H) \quad (28b)$$

can be introduced into (25) giving the simple formula

$$<O> + <H_2O>$$

$$+ 2k_{1M} [1 + \frac{k_{2M}}{k_{1M}} \cdot \frac{k_3 n(O_2)}{J_3 + a_2 n(H)}] n^2(O)$$

$$+ 2 a_5 n(OH) n(O) = 2 n(O_2) J_2 \qquad (29)$$

This shows the action of the photodissociation of oxygen and of water vapour on the aeronomic behaviour in the mesosphere. Their photodissociation rates must be known with good accuracy.

It may be added here that the production of H_2 depends strongly on the photodissociation of H_2O. Its mesospheric distribution is given by

$$<H_2> + n(H_2) [a_{24} n(O) + a^*_{H_2} n(O^*)]$$
$$= n(H) a_{23} n(HO_2) + n(H_2O) J_{H_2-O} \qquad (30a)$$

or in its simplest form

$$<H_2> = n(H_2O) J_{H_2-O} \qquad (30b)$$

i.e. a continuous production by the photodissociation of water vapour at Lyman-alpha which leads to a transport for its production peak downwards through the lower mesosphere and upwards into the thermosphere.

It should be noticed also that above the mesopause where the three-body association $H + O_2 + M$ decreases rapidly with height, the transport conditions must be considered for atomic hydrogen.

THE MESOSPHERIC PHOTODISSOCIATION OF O_2

Several difficulties occur in the determination of oxygen dissociation in the mesosphere. The first difficulty comes from the fact that the experimental absorption cross sections in the O_2 Herzberg continuum obtained before 1980 yield to an inaccurate mean value since there is a difference by a factor of 2 between the greatest and least values (see for references, Nicolet, 1983 and Bucchia et al., 1985). However, recent laboratory and theoretical determinations (Johnston, et al., 1984; Jenouvrier et al., 1986; Cheung et al., 1986; Saxon and Slanger, 1986) permits us to determine the aeronomic photodissociation of molecular oxygen between 242 and 200 nm with an acceptable accuracy (Nicolet, 1986). The various aeronomic parameters given in Table 1 lead to a correct value of the O_2 photodissociation frequency (Table II) and also of various photodissociation rates (Table III) for wavelengths greater than 200 nm. The recent O_2 absorption cross section determination are illustrated in figure 4. Thus, it can be concluded that the mesospheric O_2 photodissociation frequency is of the order of $5 \times 10^{-10} s^{-1}$ with a mean accuracy of about ± 5% in the O_2 absorption cross section and an accuracy of 5 to 10% in the solar specral irradiances. This photodissociation $J_\infty(HER) = 5 \times 10^{-10} s^{-1}$ corresponds to the mean of the extreme values indicated by Nicolet (1984), i.e.

$$J_\infty(HER) = (5 \pm 2) \times 10^{-7} s^{-1} \qquad (31)$$

with an uncertainty limit of ± 40%. There is, therefore, an improvement since it seems that both systematic and random errors are not more than ± 10 - 15%.

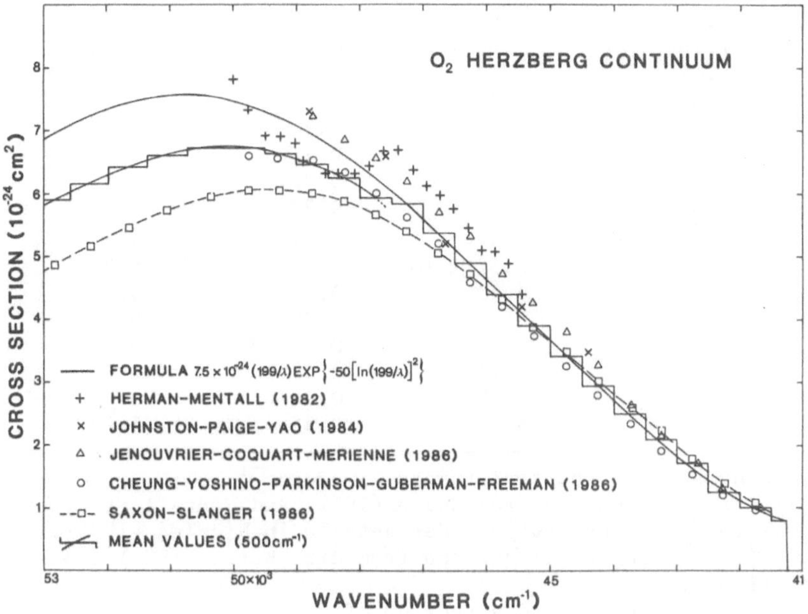

Figure 4

Recent experimental and theoretical absorption cross sections of molecular oxygen in the spectral region of the Herzberg continuum. Nicolet (1986).

Table 1. Solar spectral irradiances for $\Delta\nu$=500 cm^{-1} and corresponding attenuation cross sections in the spectral region of the Herzberg continuum (Nicolet, 1986).

$\nu(cm^{-1})^{(1)}$ $\pm$250 cm^{-1}	$q_\infty(cm^{-2}s^{-1})^{(2)}$	$\sigma(O_3)^{(3)}$ (cm^2)	$\sigma_{MS}(air)^{(4)}$ (cm^2)	$\sigma(O_2)^{(5)}$ cm^2
41250	2.04×10^{13}	9.00×10^{-18}	0.14×10^{-24}	0.80×10^{-24}
41750	1.33	7.97	0.15	1.00
42250	1.51	6.86	0.16	1.26
42750	1.26	5.79	0.17	1.73
43250	1.49	4.83	0.18	2.10
43750	1.31	4.00	0.19	2.51
44250	1.26	4.24	0.20	2.96
44750	1.52	2.55	0.21	3.42
45250	1.11	1.97	0.22	3.92
45750	1.06	1.52	0.23	4.39
46250	7.91×10^{12}	1.17	0.25	4.90
46750	8.19	8.57×10^{-19}	0.26	5.38
47250	7.12	6.50	0.27	5.84
47750	4.37	5.10	0.29	5.94
48250	2.54	4.15	0.30	6.25
48750	2.07	3.51	0.32	6.47
49250	1.69	3.18	0.34	6.63
49750	1.49	3.18	0.35	6.72

(1) Mean wave number for spectral ranges of 500 cm^{-1}.
(2) Solar irradiances adapted from Heath (1981) and Mentall et al. (1981)
(3) Ozone absorption cross section. See details in Nicolet (1986).
(4) Molecular scattering cross section from air. Bates (1984) and Nicolet (1984).
(5) Adopted mean absorption cross section of O_2 by Nicolet (1986).

Table 2. O_2 photodissociation frequency (s^{-1}) between 242 and 200 nm in the mesosphere.

Altitude (km)	sec χ = 1 J(O_2)HER	sec χ = 1.414 J(O_2)HER	sec χ = 2 J(O_2)HER	sec χ = 3 J(O_2) HER
∞	5.06×10^{-10}	5.06×10^{-10}	5.06×10^{-10}	5.06×10^{-10}
65	5.03	5.02	5.01	4.98
60	5.00	4.97	4.93	4.87
55	4.98	4.82	4.73	4.58
50	4.62	4.45	4.23	3.89

Table 3. O_2 photodissociation rates between 242 and 200 nm in the mesosphere.

Altitude (km)	sec χ = 1 (cm^{-3} s^{-1})	sec χ = 1.414 (cm^{-3} s^{-1})	sec χ = 2 (cm^{-3} s^{-1})	sec χ = 3 (cm^{-3} s^{-1})
85	1.75×10^4	1.75×10^4	1.75×10^4	1.75×10^4
80	4.40	4.40	4.40	4.40
75	9.50	9.50	9.50	9.50
70	1.92×10^5	1.92×10^5	1.92×10^5	1.92×10^5
65	3.65	3.64	3.63	3.61
60	6.64	6.61	6.56	6.48
55	1.19×10^6	1.18×10^6	1.15×10^6	1.12×10^6
50	2.07	1.99	1.89	1.74

The second difficulty in the determination of the mesospheric photodissociation of O_2 is the calculation of the absoption of the ultraviolet radiation between 175 and 200 nm in the spectral region of the Schumann-Runge bands (see, for example, Ackerman et al., 1970; Fang et al., 1974; Kockarts, 1976; Frederick and Hudson, 1979; Lewis et al., 1978 and 1979; Nicolet and Peetermans, 1980; Allen and Frederick, 1982; Yoshino et al., 1983). In our present analysis, we have adopted the reference solar spectrum adopted recently by a working group in the World Meteorological Organization, Report no. 16 entitled "Global Ozone Research and Monitoring Project" (1985).

However, we have determined a new set of transmission functions based mainly on the cross sections of Yoshino et al. (1983). The final results show that the photodissociation frequency $J_\infty(SRB)$ for the spectral range of the Schumann-Runge bands between 175 and 200 nm is

$$J_\infty(SRB) = 1 \times 10^{-7} \ s^{-1} \tag{32}$$

corresponding to the quiet sun conditions adopted by Nicolet (1984). This result indicates that there is now a general decrease in the observational results of the spectral solar irradiances. Finally, the results of a detailed calculation between $N(O_2) = 10^{19}$ cm^{-2} and $N(O_2) = 10^{22}$ cm^{-2} i.e. in the mesosphere depicted in Figure 5 for temperatures between 180 K (mesopause) and 270 K (stratopause) can be represented with a precision better than ± 10% by the formula:

$$J_{SRB}(O_2) = 3 \times 10^7/N^{0.795} \ s^{-1} \tag{33}$$

The roles of the two spectral regions 240 - 200 nm and 200 - 175 nm in the mesospheric photodissociation of O_2 are shown in Figure 6. At the mesopause (85 km), it can be seen that more than 90% of $J(O_2)$ depend on the predissociation in the Schumann-Runge bands. At the stratopause (50 km), the role of the Herzberg continuum increases since $J_{SRB}(O_2)$ reaches only 30% for an overhead sun. Nevertheless, it is important to note that these estimates based on more recent data give

$$J(O_2)_{85\ km} = 1.5 \times 10^{-8} \ s^{-1}. \tag{34}$$

for the total photodissociation frequency at 85 km for an overhead sun decreasing to

$$J(O_2)_{50\ km} = 6.75 \times 10^{-10} \ s^{-1} \tag{35}$$

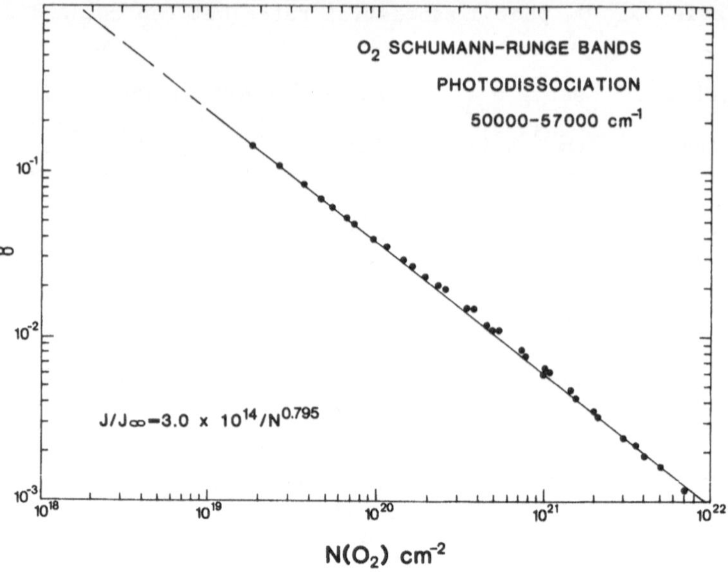

Figure 5

Relative photodissociation frequency of O_2 in the mesosphere for the spectral interval 175-200 nm corresponding to the Schumann-Runge bands. Standard solar spectral irradiances conditions and temperatures varying from 180 K (mesopause) to 270 K (stratopause).

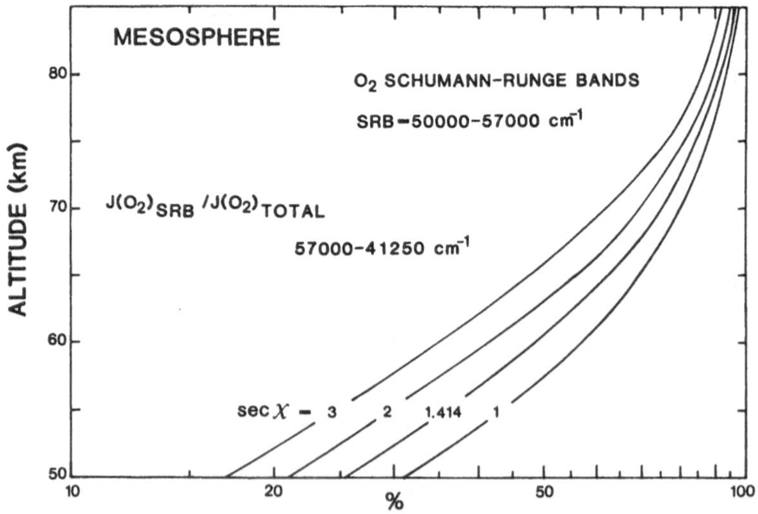

Figure 6

Ratios of the O_2 photodissociation rates in the Schumann-Runge bands interval (175-200 nm) to the total spectral range (175-242 nm) in the mesosphere.

at 50 km again for an overhead sun. These values correspond to the total O_2 photodissociation rates depicted in Figure . At the mesopause

$$2 \times 10^5 \leqslant n(O_2)J(O_2) \leqslant 5 \times 10^5 \text{ cm}^{-3} \text{ s}^{-1} \tag{36}$$

for solar zenith angles χ in the range of $3 \leqslant \sec \chi \leqslant 1$, and at the stratopause the photodissociation rate is between 2 and 3×10^6 cm^{-3} s^{-1} for the same solar conditions. The detailed values show that the Rayleigh scattering and the ozone absorption do not play any role for a total number of O_2 molecules less than 3×10^{21} cm^{-2}, i.e. at and above 60 km.

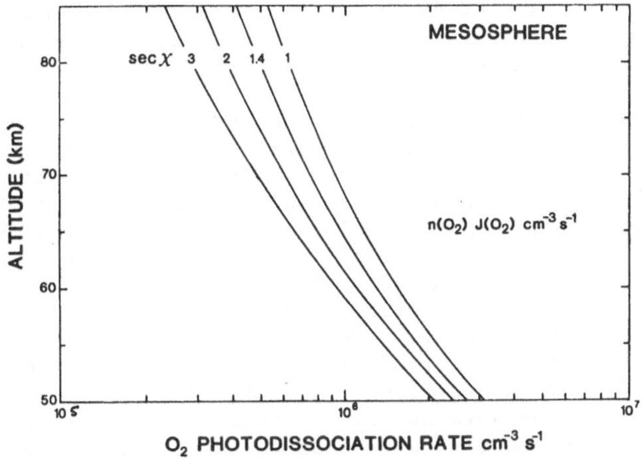

Figure 7

O_2 photodissociation rates (cm^{-3} s^{-1}) in the mesosphere for various solar zenith angles.

At 55 km, their effects for various solar angles (sec χ = 1 to 4) is always less than 2% and at the stratopause (50 km); each effect is still less than 2% for solar zenith angles less than 60° but can reach 3 to 6% for sec = χ = 3 and 4. Thus, the mesospheric transmittance is related only to the absorption of molecular oxygen.

Finally, since the atomic oxygen production between 50 and 60 km is of the order of $2 - 5 \times 10^6$ atoms cm^{-3} s^{-1}, it is clear that, in less than 10^4 seconds, the observed O_3 concentration observed in this region can be produced. It is, therefore a region of photochemical equilibrium where the aeronomic (calculated and measured) parameters should agree without the introduction of a transport effect.

Nevertheless, it is possible to consider the possibility of other dissociation processes in the mesosphere. The first additional process is related to the hydrogen Lyman-alpha radiation which reaches the middle mesosphere and will be discussed in connection with the photodissociation of water vapour. As far as the photodissociations of $^{16}O^{18}O$ molecules or of $O_2(^1\Delta_g)$, proposed by Cicerone and McCrumb (1980) and Frederick and Cicerone (1985) respectively, are concerned, they cannot constitute significant sources of atomic oxygen (see, for example, Blake et al., 1984, and Simonaitis and Leu, 1985, respectively). The $^{16}O^{18}O$ molecules cannot give more than 3 ± 1% of the $^{16}O_2$ production of oxygen atoms at mesopause levels near 70 ± 10 km. Again the absorption from $O_2(^1\Delta_g)$ is far too small to produce enough oxygen atoms.

THE SOLAR H LYMAN-ALPHA LINE

Besides causing photoionization of NO, Lyman-alpha radiation can also cause photodissociation of H_2O, CH_4, CO_2 and O_2. The experimental data obtained by Lewis et al. (1983) have been adopted by Nicolet (1985) to determine the aeronomic parameters for an analysis of the photodissociation processes in the mesosphere.

In the case of Lyman-alpha, it is necessary to take into account the variation with wavelength of the O_2 absorption cross section and of the profile of the solar line simultaneously. Both are illustrated in Figure 8. The variation of the solar H Lyman-alpha line, i.e. of tbe source function defined by $q_{i,Ly\alpha}(\Delta\lambda = 0.1A)/q_{\infty,Ly\alpha}(\Delta\lambda = 3.5\ A)$, for increasing number of O_2 molecules, is shown with its deformation for $N(O_2) = 0$, 10^{19}, 10^{20}, 2.2×10^{20}, 5×10^{20} and 10^{21} cm^{-2}, i.e. from above the mesopause to 60 km for overhead sun. The ratios $q(Ly\alpha)/q_\infty(Ly\alpha)$ are also given for the various profiles; the effective transmittance of the H Lyman-alpha irradiance varies from 0.9 for $N(O_2) = 10^{19}$cm^{-2} to 2.4×10^{-4} for $N(O_2) = 10^{21}$ cm^2, i.e. a decrease by a factor of about 10^4. For that reason the calculations have not been extended beyond an optical thickness of 10. The numerical results show that the effect of temperature (230 ± 40K) is not important for $N(O_2) < 10^{20}$ cm^{-2}. Nevertheless, the adoption of a reference temperature of 230 K, adjusted by ± 20 K for $N(O_2) > 10^{20}$ cm^{-2} leads to the following simple expression $T_{O_2}(Ly\alpha)$ with an accuracy better than ± 2%,

$$T_{O_2}(Ly\alpha) = q(Ly\alpha)/q_\infty(Ly\alpha) = \exp\left[-2.155 \times 10^{-18}\ N^{0.8855}\right] \quad (37)$$

Figure 9 shows how the effective optical depth increases from 0.1 to 10 as the total number of O_2 absorbing molecules increases from 10^{19} to 10^{21} cm^{-2}.

In aeronomic work it is necessary to know with great accuracy the absolute value of the irradiance of H Lyman-alpha and how this varies during the solar cycle. The question will not be discussed here and we will adopt 3×10^{11} photons cm^{-2} s^{-1} as a conventional value considering that the irradiance for a very quiet sun cannot be less than 2×10^{11} photons cm^{-2} s^{-1} and for an active sun cannot be more than 5×10^{11} photons cm^{-2} s^{-1}.

Since the absorption near Lyman-alpha by oxygen depends on both temperature and wavelength, the photodissociation frequency $J_{Ly\alpha}(O_2)$ defined by

$$J_{L\alpha}(O_2) = \sigma_D(O_2) q_\infty(Ly\alpha) e^{-\tau_q(O_2)} \quad (38)$$

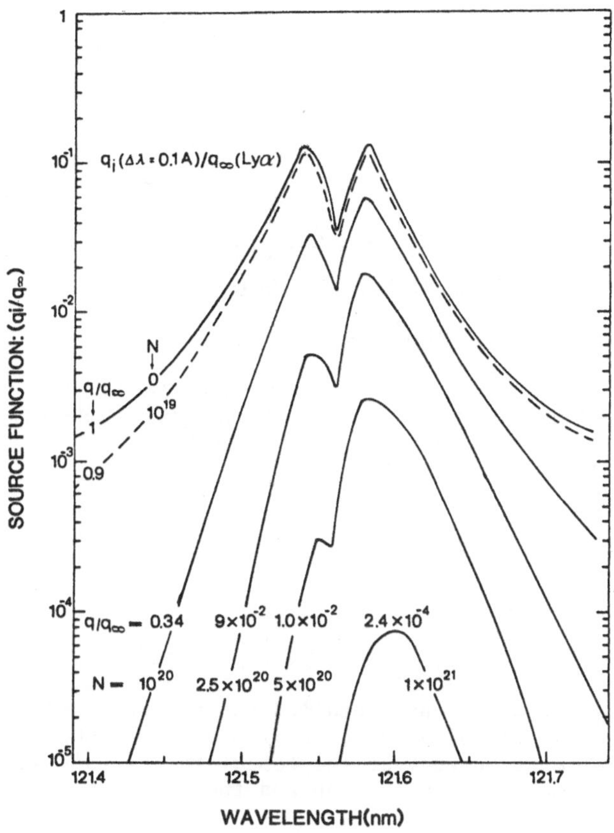

Figure 8

Variation of the profile of the solar H Lyman-alpha line with its penetration into the mesosphere. Six curves from $N(O_2)=0$ to $N(O_2) = 10^{21}$ cm^{-2} show the deformation of the profile and the corresponding decrease of the transmittance.

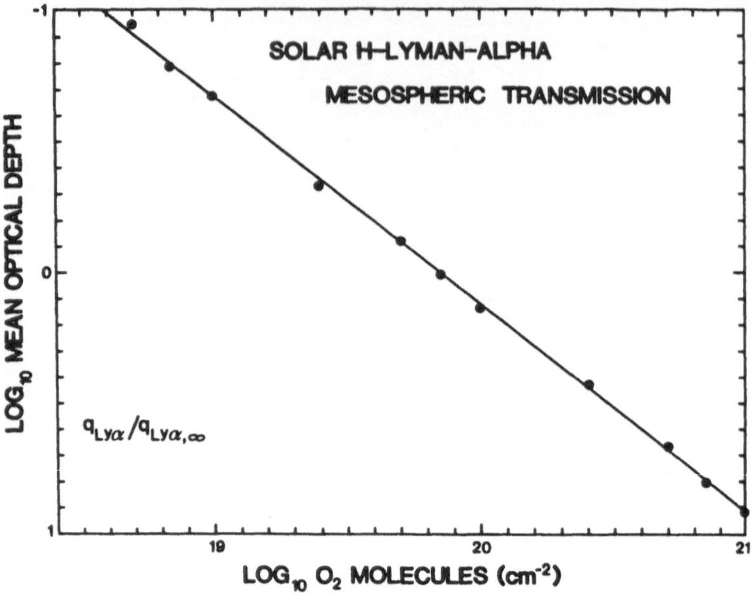

Figure 9

Mesospheric absorption of the solar H Lyman-alpha radiation represented by its effective optical depth. The straight line is obtained from the formula, equation (37), for T = 230 K.

is determining using a complete and detailed expression based on intervals of 0.1A. With the conventional value $q_\infty(\mathrm{Ly}\alpha) = 3 \times 10^{10}$ photons $\mathrm{cm}^{-2}\ \mathrm{s}^{-1}$, the O_2 photodissociation frequency at the top of the earth's atmosphere and for the mean distance between the Sun and the Earth is

$$J_{\mathrm{Ly}\alpha,\infty}(O_2) = 3.8 \times 10^{-9}\ \mathrm{s}^{-1} \tag{39}$$

and the ratio (see Figure 10) as

$$J_{\mathrm{Ly}\alpha}(O_2)/J_{\mathrm{Ly}\alpha,\infty}(O_2) = \exp\,[-4.923 \times 10^{-17}\ N^{0.8216}] \tag{40}$$

If equation (40) is used in conjunction with the result obtained in equation (23), the effective absorption-photodissociation cross-section $\sigma_D(O_2)$, (equation 38), is

$$\sigma_D(O_2)_{\mathrm{Ly}\alpha} = 1.15 \times 10^{-18}/N^{0.1175}\,\mathrm{cm}^2 \tag{41}$$

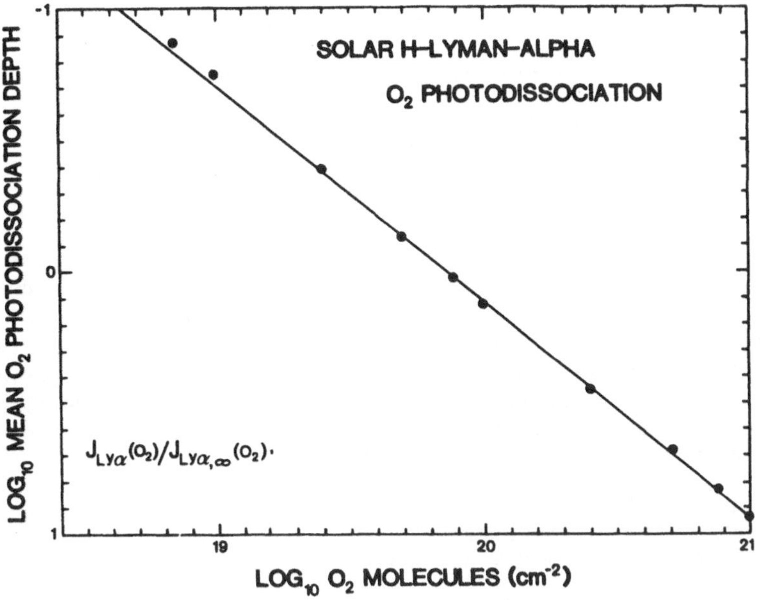

Figure 10

Equivalent optical depth of the O_2 photodissociation frequency expressed here as a "mean photodissociation depth" from 0.1 to 10 vs the total number of molecules. The straight line is obtained from formula, (equation 40), for T = 230 K.

A calculation based on these formulas shows that the O_2 photodissociation frequency due to Lyman-alpha is 10^{-9} s for $N(O_2) = 10^{20}$ cm^{-2}, i.e. increases by 25% the total photodissociation frequency between 75 km (overhead sun) and 80 km (solar zenith angle = 60°) resulting from the absorption between 175 and 242 nm. Thus, Lyman alpha plays a role in the photodissociation of O_2 in the upper mesosphere, and the photodissociation rates depicted in Figure 7 must be increased by about 25% when the Lyman alpha solar irradiance is 3×10^{11} photons $cm^{-2}s^{-1}$, and $N(O_2) = 10^{20}$ cm^{-2}.

THE PHOTODISSOCIATION OF WATER VAPOUR.

If the H_2O absorption cross section is based on the laboratory measurements of Lewis et al. (1983) with a constant mean value

$$\sigma_{D,Ly\alpha}(H_2O) = (1.57 \pm 0.05) \times 10^{-17} \text{ cm}^2 \quad (42)$$

the photodissociation of H_2O at Lyman-alpha (see successive analyses by Nicolet, 1981, 1984, 1985) can be found with a precision of about ± 3%. With the commonly used value $q_\infty(Ly\alpha) = 3 \times 10^{11}$ photons cm^{-2} s^{-1} the photodissociation frequency at the top of the earth's atmosphere is

$$J_{Ly\alpha,\infty}(H_2O) = 4.7 \times 10^{-6}\ s^{-1} \qquad (43)$$

and the mesospheric distribution of the photodissociation is, (37),

$$J_{Ly\alpha}(H_2O) = 4.7 \times 10^{-6} \exp[-2.115 \times 10^{-18}\ N^{0.8855}] \qquad (44)$$

where N is the total number of O_2 molecules determining the atmospheric transmittance.

Below the mesopause at 75 km (overhead sun) or 80 km (solar zenith angle = 60°), the photodissociation life time of water vapour due to the action of H Lyman-alpha is less than 200 hours, since $J_{Ly\alpha}(H_2O) = 1.6 \times 10^{-6}\ s^{-1}$ for $N(O_2) = 10^{20}\ cm^{-2}$. It is clear that such a short life time must produce a decrease of the H_2O mixing ratio; the reaction (15) between OH and HO_2 cannot counterbalânce production of H_2O its destruction by photodissociâtion and the transport conditions cannôt provide an adequate support.

The hydrogen production (9) related to this H_2O photodissociation by Lyman-alpha may be of the order of 10^3 H_2 molecules cm^{-3} sec^{-1} and that due to the reaction (18) H + HO_2 a value of the same order of magnitude.

The photodissociation of H_2O in the region of the Schumann-Runge bands is difficult to determine with precision since the solar flux and the H_2O absorption cross section are not known accurately enough. In addition, the transmittance of each O_2 Schumann-Runge band is very sensitive to the temperature. A general uncertainty of about a factor of 2 at the stratopause level is inherent in the various input data. We can say that, for a fixed solar spectral irradiance, it is possible to reach the following uncertainty limits: not more than ± 10% at $N(O_2) = 10^{19}\ cm^{-2}$, ± 20% at $N(O_2) = 10^{20}\ cm^{-2}$, ± 50% at $N(O_2) = 10^{21}\ cm^2$ and reaching a factor of about 2 at $N(O_2) = 10^{22}\ cm^{-2}$. A linear fit to the values of the variation of $\ln(J_{H_2O}/J_\infty)_{SRB}$ can be obtained and provides an estimation of the photodissociation of H_2O in the mesosphere. This approximation is

$$J_{SRB}(H_2O) = 2 \times 10^{-6} \exp[-3.52 \times 10^{-5}\ N^{0.2412}] \qquad (45)$$

Below the mesopause at 75 mk (overhead sun) or 80 km (solar zenith angle = 60°), the photodissociation frequency $J_{SRB}(H_2O)$ would be of the order of $1.9 \times 10^{-7}\ s^{-1}$, i.e. about 10 times less than photodissociation by Lyman-alpha.

In any case, the photodissociation at 75 km and above is the leading aeronomic process determining the vertical distribution of water vapour in the upper mesosphere and lower thermosphere. At the stratopause level, the H_2O photodissociation frequency in the 200-175 nm region is less than $10^{-8}\ s^{-1}$, and the reaction $O(^1D) + H_2O \rightarrow 2OH$ must be introduced.

Recent observations of the water vapour concentrations in the mesosphere (see for example Figure 11) described recently by Laurent et al. (1986) show that we must recognize that any estimate of the aeronomic behaviour of water vapour and its various products in the mesosphere based on available measurements and present theoretical determinations is still too uncertain. More atmospheric observations and laboratory experiments with permanent spectral solar irradiances are needed together with general improvement of the accuracy of the various aeronomic parameters.

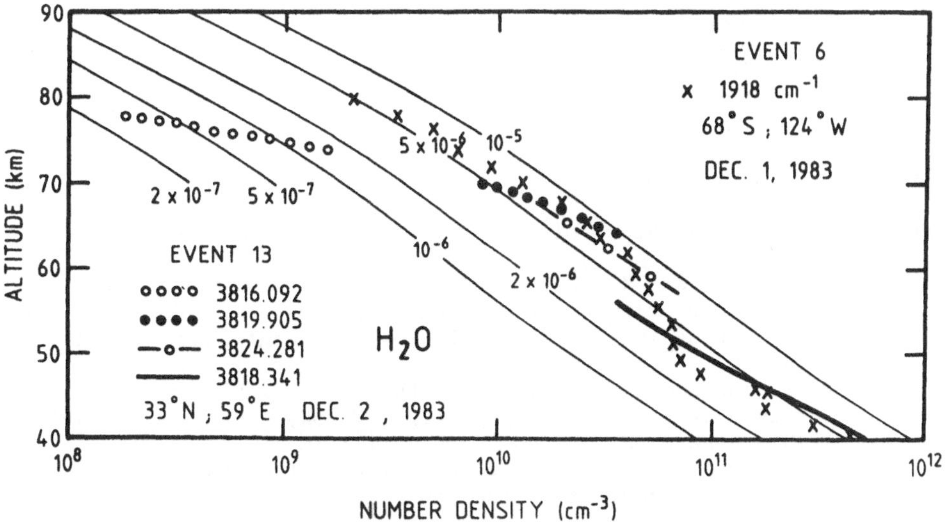

Figure 11

Observed vertical distributions of water vapour concentrations as a function of altitude at sunrise (event 6) and at sunset (event 13) by the Spacelab one grille spectrometer (see details in Laurent et al., 1986).

ACKNOWLEDGEMENTS

I wish to thank Sir David BATES for all stimulating discussions that we had during almost 40 years and I am most grateful to him for his assistance in the preparation of the final version of the manuscript.

This work was partly supported by the Direction Générale de la Science, Recherche et Développement du Programme de Recherche Environment de la Commission des Communautés Européennes.

REFERENCES

Ackerman, M., 1971, Ultraviolet solar radiation related to mesospheric processes, in Mesospheric Models and Related Experiments, edited by G. Fiocco pp. 149-159, D. Reidel, Dordrecht, Netherlands.

Ackerman , M., Biaumé, F. and Kockarts, G., 1970, Absorption cross sections of the Schumann-Runge bands of molecular oxygen, Planet. Space Sci., 18:1639.

Allen, M., and Frederick, J.E., 1982, Effective photodissociation cross sections for molecular oxygen and nitric oxide in the Schumann-Runge bands, J. Atmos. Sci., 39:2066.

Bates, D.R., 1984, Rayleigh scattering by air, Planet. Space Sci., 32:785.

Bates, D.R., and Nicolet, M., 1950a, Théorie de l'émission du spectre de la molécule OH dans le ciel nocturne, C.R. Acad. Sci. Paris, 230:1943.

Bates, D.R., and Nicolet, M., 1950b, Atmospehric hydrogen, Publ. Astron. Soc. Pacific, 62:106.

Bates, D.R., and Nicolet, M., 1950c, The photochemistry of water vapour, J. Geophys. Res., 55:301.

Blake, A.J., Gibson, S.T., and McCoy, D.G., 1984, Photodissociation of $^{16}O^{18}O$ in the atmosphere, J. Geophys. Res., 89:7277.

Bucchia, M., Megie, G., and Nicolet, M., 1985, Atmospheric transmittance and photodissociation rates in the 185-240 nm spectral range: Sensitivity of O_2 absorption cross sections in the Herzberg continuum and Schumann-Runge bands, Ann. Geophys., 3:429.

Chapman, S., 1930a, On ozone and atomic oxygen in the upper atmosphere, Philos. Mag., 10:369.

Chapman, S., 1930b, A theory of upper atmospheric ozone, Mem. Roy. Meteorol. Soc. , 3:103.

Cheung, A.S.C., Yoshino, K., Parkinson, W.H., Guberman, S.L., and Freeman, D.E., 1986, Absorption cross section measurements of oxygen in the wavelength region 195-241 nm of the Herzberg continuum, Planet. Space Sci., 34:1007.

Cicerone, R.J., and McCrumb, J.L., 1980, Photodissociation of isotopically heavy O_2 as a source of atmospheric O_3, Geophys. Res. Lett., 7:251.

Dickinson, P.H.G., Bain, W.C., Thomas, L., Williams, E.R., Jenkins, D.B., and Twiddy, N.D., 1980, The determination of the atomic concentration and associated parameters in the lower ionosphere, Proc. R. Soc. London, A 369:379.

Dickinson, P.H.G., von Zahn, U., Baker, K.D., and Jenkins, D.B., 1985, Lower thermosphere densities of N_2,O and Ar under high latitude winter conditions, J. Atmos. Terr. Phys., 47:283.

Fang, T.M., Wofsy, S.C., and Dalgarno, A., 1974, Opacity distribution functions and absorption in the Schumann-Runge bands of molecular oxygen, Planet. Space Sci., 22: 413.

Frederick, J.E., and Cicerone, R.J., 1985, Dissociation of metastable O_2 as a potential source of atmospheric odd oxygen, J. Geophys. Res., 90: 10733.

Frederick, J.E. and Hudson, R.D., 1979, Predissociation linewidths and oscillator strengths for the (2-0) to (13-0) Schumann-Runge bands of O_2, J. Molec. Spectrosc., 74:247.

Heath, D.F., 1981, A review of observational evidence for short and long term ultraviolet flux variability of the Sun, in Soleil et Climat, p. 447. Journées internationales du CNES, CNRS, DGRST, Toulouse 30 Spet; - 3 Oct. 1980.

Howlett, L.C., Baker, K.D., Megill, L.R., Show, A.W., Pendleton W.R., and Ulwick, J.C., 1980, Measurement of a structured profil of atomic oxygen in the mesosphere and lower thermosphere, J. Geophys. Res., 45:1291.

Jenouvrier, A., Coquart, A., and Merienne-Lafore, M.F., 1986, New measurements of the absorption cross sections in the Herzberg continuum of molecular oxygen in the region between 205 and 240 nm, Planet. Space Sci. 34:253.

Johnston, H.S., Paige, M., and Yao, F., 1984, Oxygen absorption cross section in the Herzberg continuum and between 206 and 327 K, J. Geophys. Res., 89:11661.

Kockarts, G., 1976, Absorption and photodissociation in the Schumann-Runge bands of molecular oxygen in the terrestrial atmosphere, Planet. Space Sci., 24:589.

Laurent, J., Brard, D., Girard, A., Camy-Peyret, C., Lippens, C., Muller, C., Vercheval, J., and Ackerman, M., 1986, Middle atmospheric water vapour observed by the Spacelab one grille spectrometer, Planet. Space Sci. 34:1067.

Lewis, B.R., Carver, J.H., Hobbs, T.I., McCoy, D.G., and Gies, H.P., 1978, Experimentally determined oscillator strengths and line widths for the Schumann-Runge band system of molecular oxygen. I. The (6-0)-(14-0) bands. J. Quant. Spectrosc. Radiat. Transfer, 20:191.

Lewis, R.R., Carver, J.H., Hobbs, T.I., McCoy, D.G. and Gies, H.P., 1979, Experimentally determined oscillator strengths and line widths for the Schumann Runge band system for molecular oxygen. II. The (2-0)-(5-0) bands, J. Quant. Spectrosc. Radiat. Transfer, 22:213.

Lewis, B.R., Vardavas, J.M., and Carver, J.H., 1983, The aeronomic photodissociation of water vapour by H. Lyman radiation, J. Geophys. Res., 88: 4935.
Mange, P., 1955, Diffusion processes in the thermosphere, Ann. Géophys., 11: 153.
Mange, P., 1957, The theory of molecular diffusion in the atmosphere, J. Geophys. Res., 62:279.
Mange, P., 1961, Diffusion in the thermosphere, Ann. Géophys., 17:277.
Meinel, A.B., 1950, OH emission bands in the spectrum of the night sky, Astrophys. J., 111: 555 and 112: 120.
Mentall, J.E., Frederick, J.E., and Herman, J.R., 1981, The solar irradiance from 200 to 330 nm, J. Geophys. Res., 86: 9881.
Nicolet, M., 1954, The aeronomic problem of oxygen dissociation, J. Atmos. Terr. Phys., 5: 132.
Nicolet, M., 1979, Etude des réactions chimiques de l'ozone dans la stratosphère, 536 pages, Edt. Institut Royal Météorologique de Belgique.
Nicolet, M., 1981a, The photodissociation of water vapour in the mesosphere, J. Geophys. Res., 86: 5203.
Nicolet, M., 1981b, The solar irradiance and its action in the atmospheric photodissociation processes, Planet. Space Sci., 29: 951.
Nicolet, M., 1983, The influence of solar radiation on atmospheric chemistry, Annales Géophys., 1: 493.
Nicolet, M., 1984a, On the photodissociation of water vapour in the mesosphere, Planet. Space Sci., 32: 871.
Nicolet, M., 1984b, On the molecular scattering in the terrestrial atmosphere: an empirical formula for its calculation in the homosphere, Planet. Space Sci., 32: 1467.
Nicolet, M., 1985, Aeronomical aspects of mesospheric photodissociation: Processes resulting from the solar H Lyman-alpha line, Planet. Space Sci., 33:69.
Nicolet, M., and Kennes, R., 1986, Aeronomic problems of the molecular oxygen photodissociation I. The O_2 Herzberg continuum, Planet. Space Sci., 34:1043.
Nicolet, M., and Mange, P., 1954, The dissociation of oxygen in the high atmosphere, J. Geophys. Res., 59:15.
Nicolet, M., and Peetermans, W., 1980, Atmospheric absorption in the O_2 Schumann-Runge band spectral range and photodissociation rates in the stratosphere and mesosphere, Planet. Space Sci., 28:85.
Offerman, D., Friedrich, V., Ross, P. and von Zahn, U., 1981, Neutral gas composition measurements between 80 - 120 km, Planet. Space Sci., 29: 747.
Saxon, R.P., and Slanger, T.G., 1986, Molecular oxygen absorption continua at 200-300 nm and O_2 radiative lifetimes, J. Geophys. Res., 91: 9877.
Simonaitis, R., and Leu, M.T., 1986, An upper limit for the absorption cross section of the oxygen $^3\Delta_u$ - $a^1\Delta_g$ transition, Geophys. Res. Lett., 12:829.
Sharp, W.E., 1980, Absolute concentrations of $O(^3P)$ in the lower thermosphere at night, Geophys. Res. Lett., 7:485.
Yoshino, K., Freeman, D.E., Esmond, J.R., and Parkinson, W.H., 1983, High absorption cross section measurements and band oscillator strengths of the (1,0) - (12,0) Schumann-Runge bands of O_2, Planet. Space Sci., 31:339.

HEAVY PARTICLE COLLISIONS : RECENT ADVANCES IN THE THEORY OF CHARGE TRANSFER BY LOW ENERGY MULTIPLY-CHARGED IONS

Ronald McCarroll*

Observatoire de l'Université de Bordeaux I (U.A. 352 du CNRS)
33270 Floirac, France

I. INTRODUCTION

The theory of collisions between heavy particles - atoms, ions, molecules - owes much to the many and varied contributions of David Bates throughout his long scientific career. The subject is broad in scope since the physical mechanisms in a collision can depend quite strongly on the specific nature of the interacting systems. This system dependence is particularly striking at low collision energies (from thermal up to around 1 keV/amu). Rather than attempt a general review of the field, I shall adopt a more pragmatic approach and concentrate on a particular class of heavy particle collisions, which will serve to illustrate the important features of recent theoretical developments. I shall be concerned with charge transfer processes involving multiply charged ions, a subject first initiated by the pioneering work of David Bates and his collaborators (Bates and Moiseiwitch, 1954 ; Dalgarno, 1957 ; Boyd and Moiseiwitch, 1957). Indeed the title of the first paper of the series entitled 'Inelastic heavy particle collisions involving the crossing of potential energy curves. I : Charge transfer from H atoms to Be^{2+}, Si^{2+} and Mg^{2+} ions' by Bates and Moiseiwitch paved the way for much subsequent work.

At that time, in the mid 1950's, the lack of experimental measurements and an unawareness of the importance of the process caused interest to wane and it was only in the late 1970's that the problem of charge transfer with multiply charged ions began to attract serious attention once more. The

*_Present address_ : Laboratoire de Physique et Optique Corpusculaires
Université Pierre et Marie Curie (Paris VI)
4, Place Jussieu, 75252 Paris Cedex

reasons are due both to experimental advances and to the recognition of the role of charge transfer as a recombination mechanism in astrophysical plasmas and in plasma fusion devices.

In the laboratory, it is now possible with sources of the ECRIS type (electron cyclotron resonance ion source) to produce ions with charge q up to 8 or more in the energy range of $\sim$ 100 eV - few keV. New recoil sources, brought recently into operation, can produce ions with energies of several eV up to a few 100 eV. Laser induced ion sources are also capable of procuding low eV multiply charged ions. New detection techniques make it possible to carry out energy gain (or loss) spectroscopy as a function of scattering angle. This wealth of experimental data has stimulated new theoretical calculations and enables us for the first time to test the validity of the theoretical models in detail.

The role of charge transfer in astrophysical plasmas was brought to attention by Péquignot et al. (1978), Péquignot (1980a, b). It was shown that in nebular type conditions - a dilute ionized gas subject to strong UV or X-radiation fields, such as planetary nebulae, HII regions of the interstellar medium, Seyfert galaxies - the reaction

$$A^{+q} + X \longrightarrow A^{+q-1} + X^{+} \qquad X = H, He \qquad 1.1$$

can become an effective recombination mechanism for multiply charged ions. When the density of neutral H or He is greater than about 10^{-3} times the electron density, account must therefore be taken of charge transfer in determining the ionization equilibria.

It may further be remarked that since charge transfer selectively populates a small number of excited states, the recombination spectra of many ions differs considerably from that produced by radiative or dielectronic recombination. There is thus a considerable need for reliable 'state-to-state' cross-sections in low energy collisions (thermal - 10 eV), where experimental data is lacking and where only theoretical predictions are available. However, since we expect the theoretical model to be valid up to energies around 1 keV/amu, where there is a great deal experimental data, a critical assessment of the theoretical results at low eV energies has now become a real possibility.

Some mention should also be made of the inverse of reaction (1.1) leading to charge transfer ionization. This can be particularly important when charge transfer recombination occurs directly via the ground state of A^{+q-1} (often the case for $q = 2$ and $X = H$ or for $q = 3$ with $X = He$). Charge transfer ionization can be quite critical for certain ions of Si and Fe (Baliunas and Butler, 1980 ; Gargaud et al., 1982).

II. PHYSICAL MODEL OF CHARGE TRANSFER

At low energies ($\lesssim$ 1 keV/amu), when the collision velocity is small compared with the velocity of an electron in a classical Bohr orbit, the charge transfer process is most simply described in terms of the molecular model.

The entry channel (A^{+q} + X) is attractive at large internuclear distances, due to the polarization potential $-\alpha q^2/2R^4$ where α is the polarizability of X and R is the internuclear distance. As a consequence, there is no activation barrier and, if conditions are favourable, the cross section may become very large at thermal energies.

For $q \geqslant 2$, the exit channels ($A^{+q-1}(n) + X^+$) are dominated by the repulsive Coulomb potential $(q-1)/R$. A curve crossing of the adiabatic curves of the $(AX)^{+q}$ molecular ion may be expected to occur at a value of R given by

$$\frac{(q-1)}{R} + \frac{q^2\alpha}{2R^4} = E_i - E_f \qquad 2.1$$

where E_i, E_f are the internal energies of the initial and final reaction products. (This formula is reasonably accurate for $R \gtrsim 5\ a_o$.) Of course, because of the Wigner non-crossing rule, most of the crossings become avoided crossings and, in general, a network of avoided crossings may be expected. A typical illustration is provided by the N^{5+}/He system (Fig. 1).

An electronic transition leading to charge transfer is only likely in the vicinity of a curve crossing, since it is only when the electronic energies are quasi-degenerate that a breakdown of the Born-Oppenheimer approximation is probable.

An approximate estimate of the collision cross sections can be deduced from the minimum energy separation at the avoided crossings by the Landau-

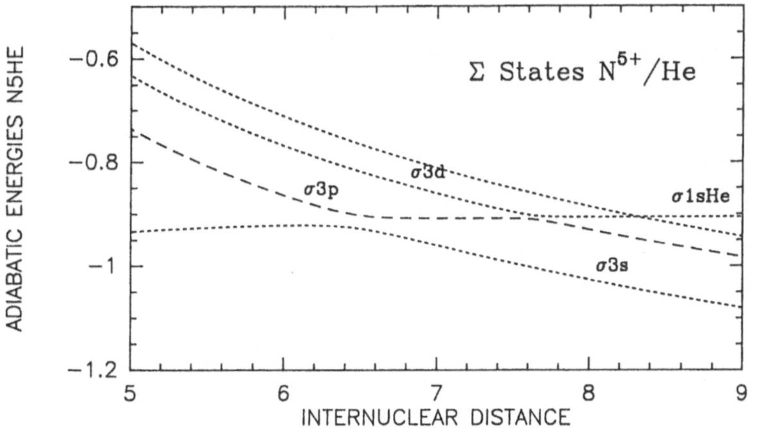

Fig. 1. Adiabatic energies of the $\sum$ states of N^{5+}/He, illustrating the avoided crossings responsible for charge transfer

Zener formula (Cf. Child, 1974). Assuming the crossing to be characterized by R_x, the position of the crossing, and Δ_x, the minimum energy separation, the cross section Q is given by

$$Q = 2\pi \int_0^{b_m} 2P(1-P)\ b\,db \tag{2.2}$$

where b_m is the maximum value of impact parameter b for which R_x is classically accessible and

$$P = \exp\left[-\frac{\pi \Delta_x^2 R_x^2}{2(q-1)v}\right] \tag{2.3}$$

where v is the radial velocity at R_x and all quantities are expressed in atomic units. For energies upwards of a few eV it is valid to assume straight-line trajectories. For thermal energies, account must be taken of possible orbiting (Langevin) effects (McCarroll and Valiron, 1976, 1979).

A transition will occur only if Δ_x is neither too small nor too large.

If $\Delta_x \gtrsim$ few eV $\quad P \to 0 \Rightarrow Q \to 0$
adiabatic limit

$\Delta_x \lesssim 10^{-2}$ eV $\quad P \to 1 \Rightarrow Q \Rightarrow 0$
diabatic limit

In practice, therefore, charge transfer is only likely to occur if

$$10^{-2} < \Delta_x < 2-3 \text{ eV}$$

Such values of Δ_x imply that R_x must be in the range

$$12a_o \gtrsim R_x \gtrsim 5a_o$$

Although for $q \geqslant 3$, there is usually at least one reaction channel for which some suitable R_x exists, the number of favourable channels is small, thereby leading to a highly selective population of excited states. Of course, the Landau-Zener model only takes account of the major primary mechanisms at avoided crossings. At energies of 500 eV/amu and upwards, the inclusion of secondary mechanisms becomes necessary.

It is also important to distinguish between two basic types of charge transfer processes (Butler and Dalgarno, 1980) :

(i) Mono-electronic capture without any change in the orbital configuration of the ionic core.

Examples

$$N^{3+}(2s^2)\,^1S + H \to N^{2+}(2s^2 3s)\,^2S + H^+ \tag{2.4}$$

$$O^{3+}(2s^2 2p)\,^2P + H \to O^{2+}(2s^2 2p3p)\,^1P,\ ^3P,\ ^1D + H^+ \tag{2.5}$$

(ii) Configuration mixing processes, where capture is accompanied by rearrangement of the ionic core orbitals.

Examples

$$N^{2+}(2s^2 2p)\,^2P + H \to N^+(2s2p^3)\,^3D + H^+ \tag{2.6}$$

$$C^{3+}(1s^2 2s)\,^2S + H \to C^{2+}(1s^2 2p^2)\,^1S + H^+ \tag{2.7}$$

III. SEMI-CLASSICAL (IMPACT PARAMETER or EIKONAL) FORMULATION OF THE COLLISION DYNAMICS

Although the application of the molecular (or perturbed stationary state) model to the collision dynamics is straightforward in principle, there are a number of fundamental difficulties linked to the incapacity of a finite molecular basis set to represent correctly the asymptotic conditions. To illustrate the point we shall first consider the problem using a semi-classical formalism, which in practice is valid at energies upwards of $\sim$ 20 eV/amu.

Let us characterize the charge transfer process by

$$(a + e)_i + b \rightarrow a + (b + e)_f \qquad 3.1$$

where a, b represent ionic cores. The electron, initially bound in state i to ion a, is captured in the collision by ion b in state f. We shall assume rectilinear trajectories defined by

$$\vec{R} = \vec{b} + \vec{v}t$$

where t is the time, $\vec{b}$ the impact parameter and $\vec{v}$ the velocity. The origin (arbitrary) of electron co-ordinates O is characterized by η as in Fig. 2. (A transformed origin O' is defined by η'.) Using the definitions of Fig. 2, we have the relations

$$\vec{r} = \vec{r}_a - \eta\vec{R} = \vec{r}\,' + (\eta' - \eta)\,\vec{R}$$

$$= \vec{r}_b + (1 - \eta)\,\vec{R} \qquad 3.3$$

The electronic Hamiltonian H may be written either as

$$H = H_a + V_a = H_b + V_b \qquad 3.4$$

where H_a is the Hamiltonian of (a + e) while H_b is the Hamiltonian of (b + e). The evolution of the system is governed by the time-dependent Schrödinger equation

$$(H - i\,\frac{\partial}{\partial t})\,\psi_{sc}(\vec{r},t) = 0 \qquad 3.5$$

We may note the importance of the choice of independent variables

$$\frac{\partial}{\partial t} = \frac{\partial}{\partial t_a} + \eta\vec{v}.\vec{\nabla}_{\vec{r}_a} = \frac{\partial}{\partial t_b} - (1-\eta)\,\vec{v}.\vec{\nabla}_{\vec{r}_b} = \frac{\partial}{\partial t'} + (\eta-\eta')\,\vec{v}.\vec{\nabla}_{r'} \qquad 3.6$$

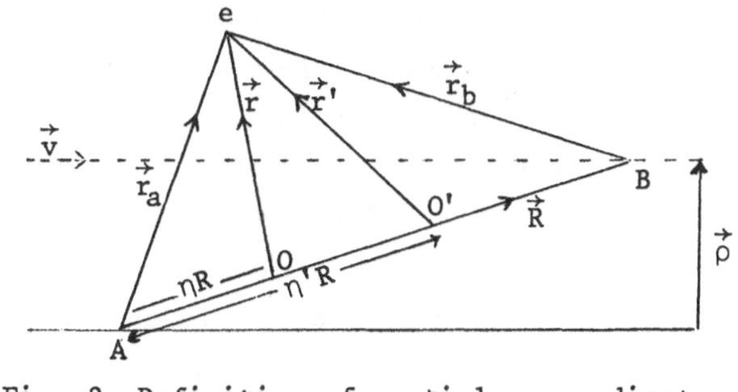

Fig. 2. Definition of particle co-ordinates

The condition of Galilean invariance implies that had we used $(\vec{r}',t)$ as independent co-ordinates rather than $(\vec{r},t)$, the wave-function $\overline{\psi}_{sc}(\vec{r}',t)$ in the $(\vec{r}',t)$ system is related to $\psi_{sc}(\vec{r},t)$ by

$$\overline{\psi}_{sc}(\vec{r}',t) = \psi_{sc}(\vec{r},t)\; e^{-i(\eta'-\eta)\vec{v}.\vec{r} + \frac{1}{2}(\eta'-\eta)v^2 t} \qquad 3.7$$

The essential idea of the molecular model is to expand $\psi_{sc}(r,t)$ on a basis set of adiabatic eigen-functions $\chi_n(\vec{r}\,;R)$

$$\psi_{sc}(\vec{r},t) = \sum_n a_n(r)\; \chi_n(\vec{r}\,;R) \qquad 3.8$$

where

$$H\,\chi_n(\vec{r}\,;R) = \epsilon_n(R)\; \chi_n(\vec{r}\,;R) \qquad 3.9$$

Herein lies the difficulty since a finite adiabatic basis set can not, strickly speaking, satisfy (3.7) and Galilean invariance is not guaranteed. Besides, matrix elements of the type

$$<\chi_n \,|\, \frac{\partial}{\partial t} \,|\, \chi_{n'}> \qquad 3.10$$

depend on η. Indeed the matrix element (3.10) may even be non-zero in the asymptotic limit, as for example when χ_n and $\chi_{n'}$ dissociate to atomic states localized on the some centre (either a or b) which are connected by an allowed dipole transition. Under these conditions, it is not even formally possible to define a unique cross-section.

Of course, in many applications, the problem is not quite so serious as might appear. From the relation

$$<\chi_n \,|\, \vec{\nabla}_r \,|\, \chi_{n'}> = (\epsilon_{n'} - \epsilon_n) <\chi_n \,|\, \vec{r} \,|\, \chi_{n'}> \qquad 3.11$$

it is clear that in the vicinity of an avoided crossing where $(\epsilon_{n'} - \epsilon_n)$ becomes small, the matrix element (3.10) will depend only weakly on η. So if the transition is localized around the avoided crossing and if excitation of the neutral target is improbable, as is the case for A^{+q}/H and A^{+q}/He collisions, a suitable choice of η can eliminate non-vanishing asymptotic couplings. This procedure is used in many applications. But there is no strict justification for such a choice. Besides, it will be unsatisfactory if both excitation and charge transfer occur simultaneously.

These ad hoc remedies have proved useful in many cases. Indeed, it is a good first step in any quantitative calculation to know how the cross section depends on η. The extent of the variation provides a measure of the failure (or the success) of the unmodified molecular model.

One possible way of overcoming the defects of the simple molecular model is to introduce 'electron translation factors', designated by ETF. The adiabatic basis set χ_n is replaced by a basis set ψ_n of the form

$$\psi_n = \chi_n \exp\left[i\, f_n(\vec{r},\vec{R})\ \vec{v}.\vec{r}\right] \qquad 3.12$$

where f_n is some suitable function depending on $\vec{r}$, $\vec{R}$ and n. The factor $\exp\left[i\, f_n\, \vec{v}.\vec{r}\right]$ is termed an electron translation factor. The form of $f_n(\vec{r},\vec{R})$ is to a large extent arbitrary, the only constraints being that

$$f_n(\vec{r},\vec{R}) \underset{R\to\infty}{\sim} \eta \qquad \text{if state n dissociates to a bound state of } (a+e) \qquad 3.13$$

$$\underset{R\to\infty}{\sim} (1-\eta) \qquad \text{if state n dissociates to a bound state of } (b+e) \qquad 3.14$$

The specific choice of f is governed by other considerations (physical, ease of application, etc.).

There are two main types of ETF's.

a) State-dependent ETF's

Of these the simplest are the plane wave type first proposed by Bates and McCarroll (1958)

$$f_n = \eta \qquad \text{if n dissociates to } (a+e) \qquad 3.15$$

$$= (1-\eta) \qquad \text{if n dissociates to } (b+e) \qquad 3.16$$

The plane wave ETF's are satisfactory at large R, when the χ_n take on their final asymptotic form. They are less satisfactory at small R when the electron is not localized around either centre also problems arise in the vicinity of an avoided crossing.

To introduce more flexibility in the choice of f_n, Crothers and Todd (1981) allow f_n to vary with R (but not $\vec{r}$). They then determine $f_n(R)$ variationnally. The shortcomings of the plane wave factors at small R can then in principle be overcome.

But the major problem associated with all state dependent ETF's arises from the complex nature of the matrix elements of the type

$$\langle \chi_n \,|\, e^{i\vec{v}.\vec{r}}\ \hat{O} \,|\, \chi_{n'} \rangle \qquad 3.17$$

where $\hat{O}$ is an operator. These integrals can be quite time consuming compared with standard molecular integrals, especially as they depend on $\vec{v}$ and ρ as well as R. Also the generalization to n-electron problems raises problems as to the symmetrisation of the ETF's.

In consequence, the use of state-dependent ETF's has been limited so far to some very simple ion-atom systems.

b) Common translation factors (CTF)

By allowing f to depend both on $\vec{r}$ and $\vec{R}$, it is possible to choose f independent of state n. For example

i) with O taken to be the mid-point of AB ($\eta = 1/2$), Schneidermann and Russek (1969) proposed

$$f(\vec{r},R) = \frac{1}{2}\ \hat{r}.\hat{R}\ \frac{1}{1+(a/R)^2} \qquad 3.18$$

where a is a constant.

ii) with O on mid-point of AB, Levy and Thorson (1969) proposed

$$f(\vec{r},\vec{R}) = \frac{1}{2}\frac{r_a^2 - r_b^2}{r_a^2 + r_b^2} \tag{3.19}$$

iii) with O on centre of charge, Vaaben and Taulbjerg (1981) proposed

$$f(\vec{r},\vec{R}) = \frac{Z_b\, r_a^3 - Z_a\, r_b^3}{2(Z_b\, r_a^3 + Z_a\, r_b^3)} + \frac{Z_a - Z_b}{2(Z_a + Z_b)} \tag{3.20}$$

iv) with O anywhere on AB (η arbitrary), Errea et al. (1984) proposed

$$f(\vec{r},\vec{R}) = \frac{R}{R^2 + \beta^2}\,\vec{r}.\hat{R} \tag{3.21}$$

Note that if $\beta = 0$, this form of f preserves Galilean invariance.

As $R \to \infty$, the electron becomes localized either on centre a or b, and f tends to the required limit. The CTF's are much simpler to apply than state dependent ETF's since the matrix elements involved reduce to standard molecular integrals. But, of course, there is no guarantee that any particular form of f is either reasonable or optimal.

As an example of a CTF, let us consider (3.21) with $\beta = 0$. The radial matrix element $<\chi_n \mid \partial/\partial R \mid \chi_{n'}>$ of the molecular model is replaced by

$$<\chi_n \mid \partial/\partial R - \frac{1}{R}\, z\, \frac{\partial}{\partial z} \mid \chi_{n'}> \tag{3.22}$$

in the CTF model. It is easily verified that as $R \to \infty$, this matrix element will always vanish.

However, although the introduction of CTF's remove the most serious drawbacks of the molecular model, difficulties still remain. Let us, for example, consider a translating atom (or for example the collision of an atom with a non-interacting particle such as a neutron). The adiabatic states are simply atomic states of (a + e) which we designate by $\mid \phi_n^a >$. The matrix element (3.22) then reduces to

$$\frac{1}{R} < \phi_n^a \mid z_a \frac{\partial}{\partial z_a} \mid \phi_{n'}^a > \tag{3.23}$$

If n, n' are of the some parity (3.23) will be non-vanishing for finite R. As a consequence, there will be a non-vanishing probability for excitation of (a + e). The probability is small in most practical cases but it should be identically zero. We may remark that the state-dependent ETF's give the correct result for this problem.

It may be concluded that there is at present no definitive solution to the ETF problem. However, as we shall see in later applications, the use of CTF's often provides an acceptable solution.

Of course, if we abandon the use of adiabatic states, other approaches are possible. Thus, Fritsch and Lin (1982) use a large basis set of atomic orbitals (with plane wave ETF's). Their basis includes united atom orbitals which allow for a good representation of the collision in the molecular region. However, it is then no longer possible to interpret the results in terms of curve-crossings. In the approach of Lüdde and Dreizler (1981), the use of ETF's is avoided altogether by using a very large basis set of spheroidal-type orbitals. The method has proved satisfactory for some simple cases but the generalization to complex systems would appear difficult. Winter and Lane (1985), and Kimura and Lin (1985) have proposed a mixed atomic and molecular state expansion - molecular model in the interaction region $R \geqslant R_0$. The results are promising but the optimisation of the matching radius R_0 is not altogether satisfactory.

However, in the present state of the art, it would seem that the introduction of CTF's present the most practical solution to the problem at least in the low energy regime.

IV. QUANTUM MECHANICAL FORMULATION

When the energy transfer is comparable with the collision energy, a quantum mechanical formulation is preferable. However, the direct introduction of ETF's in a quantal formulation is inconvenient and an alternative way to taking account of translation effects is required. The simplest way is to introduce suitable 'reaction co-ordinates', which adapt themselves automatically to the correct asymptotic conditions (Mittleman, 1969 ; Thorson and Delos, 1978 ; Delos, 1981 ; Soloviev and Vinitsky, 1985). As we shall see the notion of reaction co-ordinates in a quantal formulation is analogous to the CTF's of the semi-classical method.

Let us recall the 3 Jacobi co-ordinate systems for the 3-body problem, which allow us to express the kinetic energy operator T in separable form

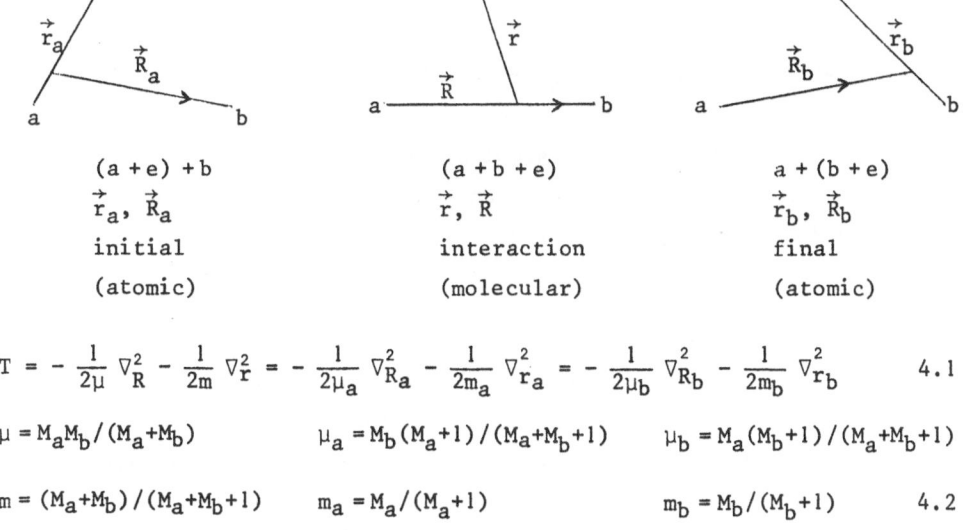

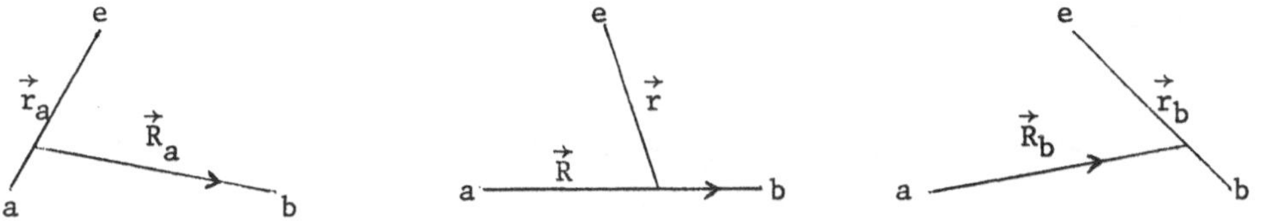

(a + e) + b	(a + b + e)	a + (b + e)
$\vec{r}_a$, $\vec{R}_a$	$\vec{r}$, $\vec{R}$	$\vec{r}_b$, $\vec{R}_b$
initial	interaction	final
(atomic)	(molecular)	(atomic)

$$T = -\frac{1}{2\mu}\nabla^2_R - \frac{1}{2m}\nabla^2_r = -\frac{1}{2\mu_a}\nabla^2_{R_a} - \frac{1}{2m_a}\nabla^2_{r_a} = -\frac{1}{2\mu_b}\nabla^2_{R_b} - \frac{1}{2m_b}\nabla^2_{r_b} \qquad 4.1$$

$$\mu = M_aM_b/(M_a+M_b) \qquad \mu_a = M_b(M_a+1)/(M_a+M_b+1) \qquad \mu_b = M_a(M_b+1)/(M_a+M_b+1)$$

$$m = (M_a+M_b)/(M_a+M_b+1) \qquad m_a = M_a/(M_a+1) \qquad m_b = M_b/(M_b+1) \qquad 4.2$$

The wave-function of the initial state $(a+e)_i$ may then be expressed as

$$\exp(i\vec{k}_i \cdot \vec{R}_a)\ \phi_i(\vec{r}_a) \qquad 4.3$$

while that of the final state $a+(b+e)_f$ has the form

$$\exp(i\vec{k}_f \cdot \vec{R}_b)\ \phi_f(\vec{r}_b) \qquad 4.4$$

where $\vec{k}_i$, $\vec{k}_f$ are the initial and final wave vectors.

The co-ordinates $(\vec{r}_a, \vec{R}_a)$ are required to describe the initial state whereas $(\vec{r}_b, \vec{R}_b)$ are required for the final state. So no Jacobi co-ordinate system is adequate to describe the entire collision process. For example, if we use (as is frequently done) the $(\vec{r}, \vec{R})$ co-ordinate system and expand the total wave-function ψ in the form

$$\psi(\vec{r}, \vec{R}) = \sum \vec{F}_n(\vec{R})\ \chi_n(\vec{r}\,;R) \qquad 4.5$$

we obtain the following set of coupled differential equations (in matrix notation)

$$-\frac{1}{2\mu}\left[\nabla_R^2 + 2i\ \vec{\underline{P}}.\vec{\nabla}_R + i\ \vec{\nabla}_R.\vec{\underline{P}} + \underline{P}^2\right]\underline{F}(\vec{R}) = E\ \underline{F}(\vec{R}) \qquad 4.6$$

where $\vec{P}_{mn} = \langle \chi_m \,|\, -i\ \vec{\nabla}_R \,|\, \chi_n \rangle$ 4.7

Note that $\vec{\nabla}_R$ must be carried out with respect to $\vec{r}$ fixed. As in the semi-classical formulation, the matrix elements $\vec{P}_{mn}$ may be non-vanishing in the asymptotic limit. Similar problems occur had we used $(\vec{r}_a, \vec{R}_a)$ or $(\vec{r}_b, \vec{R}_b)$ for our co-ordinate system. To overcome the problem, one possibility is to introduce a co-ordinate system depending in the position of the electron. For example, introducing the variables $(\vec{u}, \vec{r})$ where

$$\vec{u} = \vec{R} + \frac{1}{\mu}\vec{s} \qquad 4.8$$

$$\vec{s} = \frac{\vec{r}.\vec{R}}{R^2}\left(\vec{r} - \frac{1}{2}\frac{\vec{r}.\vec{R}}{R^2}\vec{R}\right) \qquad 4.9$$

it is easy to verify that $\vec{u}$ tends to $\vec{R}_a$ or $\vec{R}_b$ depending on whether $R/r_a \to \infty$ or $R/r_b \to \infty$. So the co-ordinates $(\vec{u}, \vec{r})$ can be used to describe both the initial and final asymptotic conditions. Of course the choice of $\vec{s}$ given by (4.9) is not unique. It is the simplest choice which eliminates artificial isotopic effects (Galilean invariance in semi-classical language) and guarantees the correct asymptotic form.

The full wave function is now expanded as

$$\psi(\vec{u}, \vec{r}) = \sum T_n(\vec{u})\chi_n(\vec{r}\,;\vec{u}) \qquad 4.10$$

where the adiabatic states are defined with respect to $\vec{u}$ constant. The justification is that the adiabatic functions will not change appreciably if we replace R by u whatever the value of $\vec{r}$ (consequence of the small mass ratio $1/\mu$). A straightforward (but lengthy) calculation yields

$$-\frac{1}{2\mu}\left[\nabla_u^2 + 2i(\vec{\underline{P}}+\vec{\underline{A}}).\vec{\nabla}_u + i\,[\vec{\nabla}_u.(\vec{\underline{P}}+\vec{\underline{A}})] + (\vec{\underline{P}}+\vec{\underline{A}})^2\right]\underline{T}(\vec{u}) = E\,\underline{T}(\vec{u}) \qquad 4.11$$

where $A_{mn} = i\,(\epsilon_m - \epsilon_n) < \chi_m \mid \vec{s} \mid \chi_n >$ 4.12

and $\vec{P}_{mn}$ is as defined by (4.7). It is easy to verify that the matrix elements of $(\vec{\underline{P}}+\vec{\underline{A}})$ vanish in the asymptotic limit.

A partial wave decomposition can be carried out in the usual way (bearing in mind that the adiabatic wave-functions are defined with respect to body-fixed co-ordinates). We obtain two types of coupling matrix elements

- radial coupling elements of the form $<\chi_m \mid \frac{\partial}{\partial R} + \frac{z}{R}\frac{\partial}{\partial z} \mid \chi_n>$ 4.13

- rotational coupling elements of the form $<\chi_m \mid i\,L_y + (x\frac{\partial}{\partial z} + z\frac{\partial}{\partial x}) \mid \chi_n>$ 4.14

where the z-axis is in the direction of $\vec{R}$ and the y-axis is perpendicular to the classical collision plane.

These are identical to the matrix elements arising in the CTF semi-classical formula. The close association of reaction co-ordinates and CTF's is thus established.

V. SOLUTION OF COUPLED EQUATIONS IN THE QUANTUM MECHANICAL FORMALISM AND INTRODUCTION OF DIABATIC STATES

To solve (4.11), we expand $T_n(\vec{u})$ on a basis set of symmetric top functions $D^K_{\Lambda_n M}$ where K is the total angular momentum, M its projection on the z-axis and Λ_n the projection of the angular momentum of molecular state n on the internuclear axis.

$$T_n(\vec{u}) = \sum_{KM} (-)^K \{(2K+1)/4\pi\}^{1/2}\, D^K_{\Lambda_n M}\; f_n^{(K)}(u)\,\frac{1}{u} \qquad 5.1$$

which yields

$$\frac{d^2}{du^2}\underline{f}^{(K)} + 2\underline{B}\frac{d}{du}\underline{f}^{(K)} + \underline{W}\,\underline{f}^{(K)} = 0 \qquad 5.2$$

where

$$B_{ij} = <\chi_i \mid \frac{\partial}{\partial R} \mid \chi_j>\,\delta(\Lambda_i\,,\Lambda_j) \qquad 5.3$$

$$W_{ij} = V_{ij} + C_{ij} + \frac{2}{u^2}K\,L_{ij}\,\delta(\Lambda_i,\,\Lambda_j \pm 1) \qquad 5.4$$

$$V_{ij} = \left[2\mu(E-\epsilon_i) - \frac{K(K+1)}{R^2}\right]\delta(i\,,j) \qquad 5.5$$

$$C_{ij} = <\chi_i \mid \frac{\partial^2}{\partial R^2} \mid \chi_j> \qquad L_{ij} = <\chi_i \mid i\,Ly \mid \chi_j> \qquad 5.6$$

To solve (5.2) it is convenient to remove the first order derivative by the transformation

$$\underline{g}^{(K)} = \underline{D}\,\underline{f}^{(K)} \qquad 5.7$$

where

$$\frac{d}{dR}\underline{D} + \underline{B}\,\underline{D} = 0 \qquad\qquad \underline{D}(\infty) = I \tag{5.8}$$

giving

$$\frac{d^2}{dR^2}\underline{g}^{(K)} - 2\mu\,\underline{V}^d\,\underline{g}^{(K)} + \left\{2\mu\,E - \frac{K(K+1)}{R^2}\right\}\underline{g}^{(K)} = 0 \tag{5.9}$$

where

$$\underline{V}^d = \underline{D}^{-1}\left[\underline{\epsilon} - \frac{1}{2\mu\,u^2}\,K\,\underline{L}\right]\underline{D} \tag{5.10}$$

is termed the diabatic matrix (as originally defined by Smith (1969)). Although the diagonal elements of $\underline{V}^d$ usually lie close to the adiabatic energies with the avoided crossings replaced by real crossings, a direct physical interpretation of $\underline{V}^d$ can be misleading if the non-diagonal terms become large. As a general rule, it is safest to consider the introduction of $\underline{V}^d$ as a useful mathematical transformation to solve the coupled equations.

A typical example of a diabatic transformation is illustrated in Fig. 3 - 5 for the system C^{4+}/H. It may be noted that both the diagonal and off diagonal elements of $\underline{V}^d$ vary smoothly in contrast with the matrix elements of $\underline{\vec{P}}$.

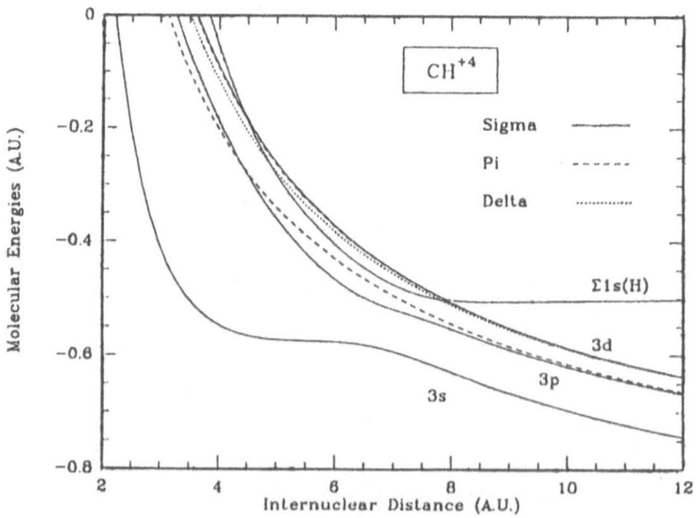

Fig. 3. Potential energy curves of CH^{4+}. The solid curves refer to Σ states, the dashed curve to π states and the dotted curve to Δ states.

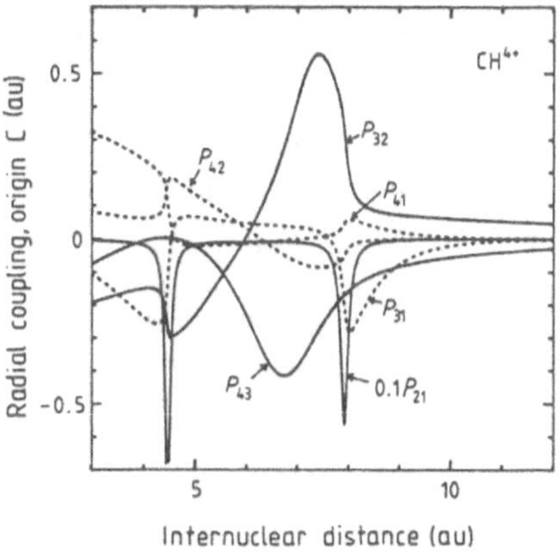

Fig. 4. Radial coupling matrix elements between the Σ states of CH^{4+}

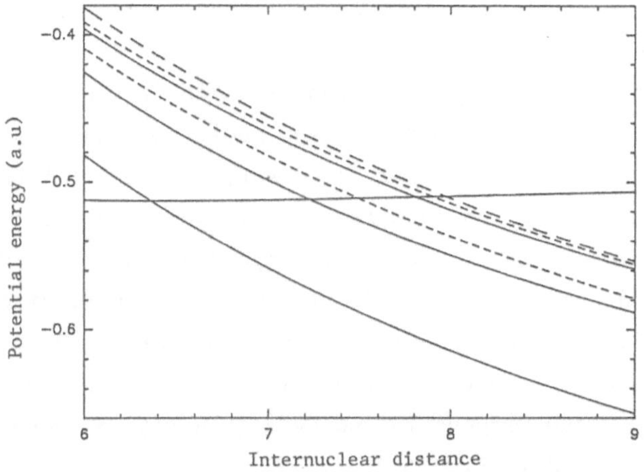

Fig. 5. Diagonal elements of the diabatic matrix V^d. Solid curves refer to Σ states, short dashed curves to π states and long dashed curves to Δ states

VI. CASE STUDIES

From the preceding analysis, it is clear that since the location of the avoided (and real) potential energy curve crossings may vary widely from one system to another, a general classification of charge transfer reactions is quite difficult. It is therefore instructive to consider in detail a few representative systems, which illustrate the main features of the theoretical model.

VI.1. System C^{3+}/H

The C^{3+}/H system furnishes an interesting (and typical) example of the critical role of translation effects in the diabatic regime. This system has been the subject of much experimental investigation - translational energy spectroscopy (Wilkie et al., 1986), photon emission spectroscopy (Ciric et al., 1985), total charge transfer (Phaneuf et al., 1982).

In the molecular model, we may consider separately the singlet and triplet reaction products. In the singlet case, the most likely reactions are type II processes

$$C^{3+}(1s^2 2s) + H \rightarrow C^{2+}(1s^2 2p^2)\,^1S + H^+ \qquad R_X = 4.7\ a_o$$

$$\rightarrow C^{2+}(1s^2 2p^2)\,^1D + H^+ \qquad R_X = 3.4\ a_o$$

while in the triplet case, the dominant channels are type I processes

$$C^{3+}(1s^2 2s) + H \rightarrow C^{2+}(1s^2 2s3s)\,^3S + H^+ \qquad R_X = 11.5\ a_o$$

$$\rightarrow C^{2+}(1s^2 2s3p)\,^3P + H^+ \qquad R_X = 26$$

Let us confine ourselves to the triplet case. Here we have an isolated long-distance crossing, a situation quite common for ions with charge $q = 3$ or 5. This crossing tends to be diabatic, so that as the energy increases the transition tends to occur somewhat inwards of the crossing. The LZ model proves to unreliable under these conditions and detailed semi-classical or quantal calculations are necessary. Moreover, translation effects are considerably amplified when the transition is not localized in the vicinity of the crossing. (The origin dependence of the radial matrix elements is minimal at R_X.) The C^{3+}/H system therefore provides an interesting test case for the translation factor problem (Opradolce et al., 1987).

The calculated molecular potential energies and the radial coupling matrix both with and without including a translation factor of the CTF type are illustrated in Fig. 6 and 7.

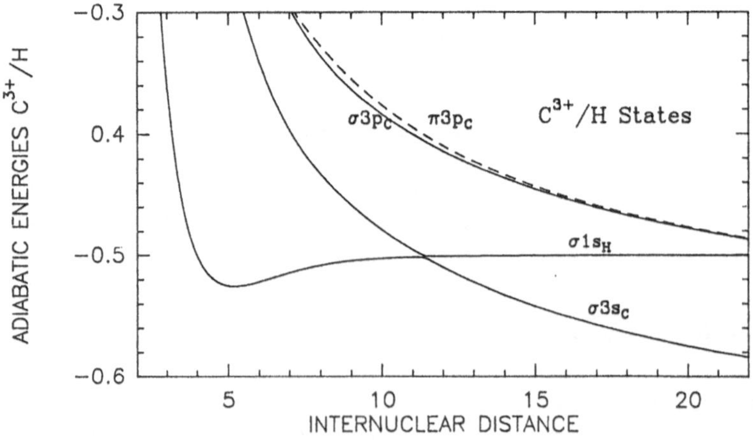

Fig. 6. Potential energy curves of the triplet states for C^{3+}/H

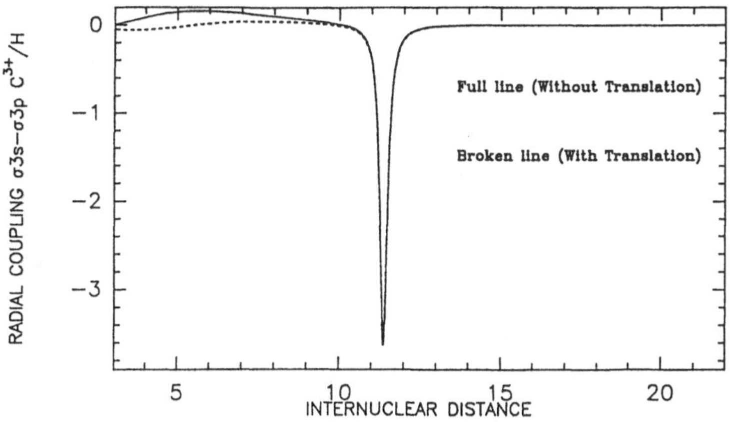

Fig. 7. Radial coupling matrix elements for capture into the $(2s3s)\,^3S$ state of C^{2+}

It is immediately observed that the inclusion of CTF's only modifies the coupling at internuclear distances, well inwards from the crossing. However, the results of a 2-state semi-classical calculation show that the translation effect is quite considerable even for energies as low as 100 eV/amu. The results are tabulated below.

Cross-sections (in units of $10^{-16}\,cm^2$) for capture into the $(1s^2 2s3s)\,^3S$ state of C^{2+} in C^{3+}/H collisions. Results of 2-state model calculation

E (eV/amu)	Origin on H	Origin on C	C T F
30	4.47	4.97	4.70
75	3.26	4.12	3.65
125	2.85	4.10	3.39
250	2.70	4.97	3.60
500	2.88	6.55	4.23
1000	2.95	9.44	5.09
2000	3.24	13.15	6.26

More exhaustive calculations taking account of coupling with the Σ and π states correlated to the $C^{2+}(1s^2 2s2p)\,^3P$ channel do not modify these conclusions, even though secondary coupling mechanisms lead to an appreciable population of the ^3P state when the collision energy exceeds 1 keV/amu. The theoretical CTF results lie within the experimental error bars and would indicate that translation effects have been correctly accounted for in our model.

It is thus seen that accurate molecular structure calculations alone do not suffice to predict reliable cross sections in the diabatic regime, which for C^{3+}/H occurs at energies upwards of 30 eV/amu. This result is quite typical for systems where the primary charge transfer mechanism takes place via a long-distance avoided crossing. The neglect of translation effects by Bienstock et al. (1982) would therefore seem unjustified.

VI.2. System C^{4+}/H

The C^{4+}/H system provides an interesting test of many aspects of the theory. Although the LZ model describes the qualitative features of the charge transfer reaction, it does not succeed in reproducing many of the finer details to adequate accuracy. Translation effects, though less striking than for C^{3+}/H, can have an appreciable effect at keV energies and the role of rotational coupling interferes with the radial coupling mechanism in certain reaction channels.

The two tightly-bound electrons of the ^1S C^{4+} core only play a passive role in the collision and it is possible to treat the molecular structure as an effective one electron problem (Gargaud et al., 1981). In contrast with completely stripped ions, the sub-ℓ levels of the configuration C^{3+} $(1s^2n\ell)$ are not degenerate (although they do lie close to one another), which facilitates the comparison between theory and experiment. Moreover, the presence of a non-coulombic ionic core has a profound effect on the capture process at low eV energies. A detailed study of a system such as C^{4+}/H can help us to understand why charge transfer cross sections of the type A^{4+}/H can cary very greatly from one type of ion to another.

The molecular energies of CH^{4+}, calculated by a model potential method, have been already given in Fig. 3. Three well defined avoided crossings occur involving capture into the $(1s^2n\ell)$ configuration of C^{3+} with $n\ell = 3d$, 3p, 3s. The crossings and minimum energy separations are given below.

$n\ell$	$R_x(a_0)$	$\Delta_x(eV)$
3d	7.9	0.093
3p	7.4	1.09
3s	7.1	2.01

A Landau-Zener (LZ) calculation indicates that we may expect capture into the 3d state to dominate at low energies ($\leqslant$ 3 eV/amu), whereas at higher energies capture into the 3 p state dominates. Capture into the 3s state will become increasingly important at energies exceeding several 100 eV/amu.

These previsions are confirmed by the detailed calculations (either quantum mechanical or semi-classical) of Gargaud et al. (1981), Gargaud and

Total charge transfer cross-sections (in units of 10^{-16} cm^2) in C^{4+}/H collisions

E eV/amu	4-state (radial only)		7-state (radial + rotational)	
	Origin on C nucleus	CTF	Origin on C nucleus	CTF
0.9	14.1	14.1	14.1	14.1
8.8	7.6	7.6	8.1	8.2
29	14.8	15.0	17.8	18.1
59		19.5		26.1
88	21.9	22.5	31.3	31.8
147	24.8	25.5	36.7	37.8
590	36.4	38.7	38.4	39.4
1180	35.8	38.7	38.4	39.4

McCarroll (1985). But the LZ model takes account only of the primary capture mechanisms at the avoided crossings. The detailed calculations take account also of rotational coupling and of translation effects. Rotational coupling between the $\Sigma 3p$ and $\pi 3p$ adiabatic states is found to contribute considerably to the primary mechanism. Translation effects are also appreciable but not of major importance for the total charge transfer cross-section. Some selected results are presented above.

Although there is good agreement between theory and the experiments of Phaneuf et al. (1982) at energies up to 50 eV/amu, the theoretical results lie considerably above experiment in the 100 eV - 1 keV/amu range (see Fig. 8).

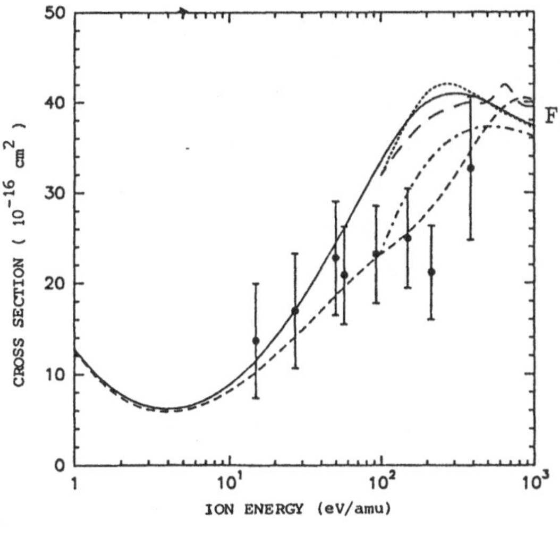

Fig. 8. Total charge transfer cross-sections in C^{4+}/H collisions. The experimental data of Phaneuf et al. is compared to the 7-state CTF (solid curve) and 4-state CTF (dashed curve) results of Gargaud et al. (1987). The theoretical results of Fritsch and Lin (1984) and Hanssen et al. (1984) are given at energies above 100 eV/amu

	E keV/amu	P_{3s}	P_{3p}	P_{3d}
4-state origin C	1.18	.34	.50	.16
4-state CTF	1.18	.41	.45	.14
7-state origin C	1.18	.27	.46	.27
7-state CTF	1.18	.35	.41	.24
Expt (Dijkkamp et al., 1985)	1.11	.34	.41	.25
Fritsch and Lin (theory)	1.0	.34	.42	.24

The most detailed comparison between theory and experiment is provided by the branching ratios $P_{n\ell}$ for capture into state $n\ell$. The results of our calculations, with and without CTF's, are summarized above.

The 4-state calculations take account only of radial coupling between Σ states, while the 7-state calculations take account of both radial and rotational coupling between the full manifold of states correlated to the reaction products. It may be noted that only the 7-state CTF calculations are in good accord with experiment, which underlines the importance of both rotational and translation effects. In view of this well-nigh perfect agreement, it is difficult to understand the discrepancy with Phaneuf et al. for the total cross sections.

Several delicate points showed up in the course of the calculations :

- sensitivity of the rotational coupling effect to small changes in the energy of the $\pi 3p$ state ;
- sensitivity of the 3d capture cross section at low eV energies to the energy spacing of the 3p, 3d states of C^{3+}.

Let us examine these points in turn.

a) Influence of rotational coupling. The results of the 7-state (4 Σ states, 2 π states, 1 Δ state) calculation indicate that for energies between 30 and 500 eV/amu, electron capture into the 3p state dominates. The main rotational coupling mechanism is between the $\Sigma 3p$ and $\pi 3p$ states.

A simplified 3-state model has been studied by Gargaud et al. (1987). There are two main couplings - diabatic Σ-Σ coupling H_{12} and the rotational Σ-π rotational coupling H_{23} (see Fig. 9). The rotational coupling H_{13} is negligible. The energy of the $\pi 3p$ state is varied according to

$$\epsilon_3 = \epsilon_2 + \Delta\epsilon \qquad 6.1$$

where $\Delta\epsilon$ is a small arbitrary constant. The coupling H_{13} is not affected by small variations of $\Delta\epsilon$.

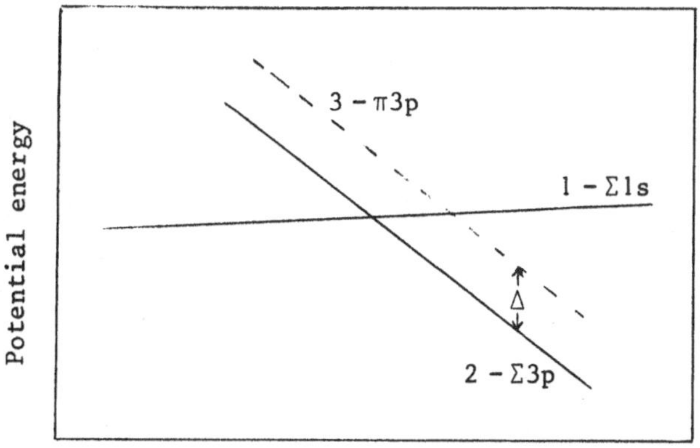

Fig. 9. Schematic diabatic states in the 3-state model

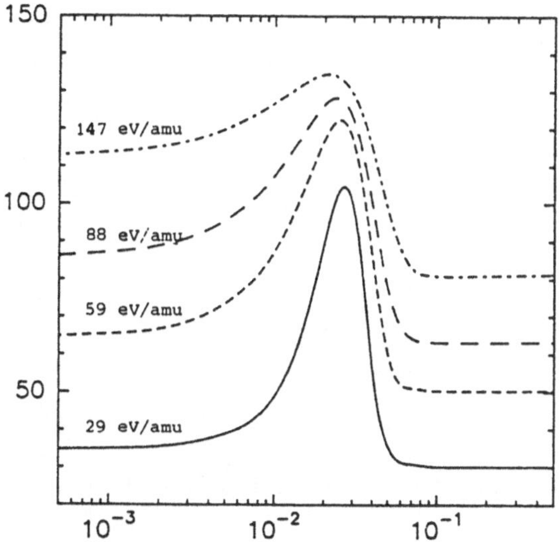

Fig. 10. Results of 3-state model plotted as a function of the energy splitting of the Σ3p-π3p diabatic states

The results, presented in Fig. 10, illustrate how the total charge transfer cross-section depends on Δ. It is seen that the rotational coupling effect is maximum for Δ around 0.6 eV. Our molecular structure calculations yield Δ∿0.4 eV in the vicinity of the crossing. The charge transfer cross section is then very sensitive to quite small inaccuracies in the Σ-π energy separations at long distances. This point often seems to have been overlooked.

b) Influence of the 3p - 3d energy separation. Since molecular structure calculations are always liable to be some error, it is important to estimate the consequences of such errors on the cross sections.

	E_{3s} (eV)	E_{3p} (eV)	E_{3d} (eV)
M1	26.956 (0.08 %)	24.813 (0.05 %)	24.193 (0.05 %)
M2	26.932 (0.01 %)	24.936 (0.5 %)	24.211 (0.03 %)
M3	26.935 (0 %)	24.800 (0 %)	24.204 (0 %)

We have used different model potentials M1, M2, M3 for the C^{4+} core which reproduce the excited states of C^{3+} to varying degrees of accuracy (Gargaud and McCarroll, 1986). The calculated energies of the 3s, 3p, 3d states (and the percentage error) are given in the table above.

All 3 models lead to an accurate value of R_x, but only M1 and M3 give good energy separations. The error in $(E_{3d} - E_{3p})$ is important in the molecular calculation on Δ_x, which is the critical parameter in the cross-section determination. The results, presented below, show that M1 and M3 give good values of Δ_x but M2 is in error by almost 40 %.

	$(E_{3d}-E_{3p})$ (eV)	R_x (a_o)	Δ_x (eV)
M1	0.620 (4 %)	7.92	0.102
M2	0.725 (21.6 %)	7.88	0.138
M3	0.596 (0 %)	7.90	0.093

As a result M2 leads to an overestimation of the capture cross-section (by as much as a factor of 2 at energies $\leqslant$ 15 eV/amu). We have checked, using different model potentials, that the critical parameter is indeed the 3d-3p energy separation. The reason would appear to arise from a partial overlapping of the 3d and 3p avoided crossings. This is in contrast to the case of an isolated crossing where this extreme sensitivity is not observed.

In spite of its apparent simplicity, the C^{4+}/H exhibits many surprising complications. We have seen how essential it is to have very accurate potential energy curves at long distances. The agreement of theory with Dijkkamp et al. is indeed gratifying since little improvement of the theoretical model can be expected by the inclusion of other molecular states. On the other hand, the results of Phaneuf et al. are difficult to understand for if our molecular structure calculations were in error, the disagreement with Phaneuf would be greatest at low energies. As to the influence of rotational coupling on capture into the 3p state, we may expect this to be typical of systems (for example Al^{3+}/H) where capture occurs into a state of non zero angular momentum ($l \neq 0$) via an avoided crossing in the adiabatic regime.

VI.3. Charge transfer of Ar^{6+} and Ar^{8+} with He

Although from the theorist's viewpoint, charge transfer with atomic H is relatively simple to investigate, experimentally it is easier to study charge transfer with chemically stable targets such as He or H_2 and the data from such experiments tend to be more reliable. Besides, the ionization potential of He being greater than for H, the quantum levels populated for a given ionic charge are lower than for H, so that even for ions with a charge up to ~ 8, the number of effective reaction channels is still within manageable proportions in the molecular model.

Two particularly interesting systems are Ar^{6+}/He and Ar^{8+}/He, which have been studied experimentally both at keV energies by conventional ion sources (Panov, 1981) and at low energies by recoil ion techniques (Justiniano et al., 1984 ; Hvelplund et al., 1986 ; McCullough et al., 1986). These systems exhibit typical avoided crossings at large internuclear distances.

The ionic Ar^{6+} and Ar^{8+} cores being more diffuse than the C^{4+} core, the sub-ℓ levels are more widely separated than in C^{3+} and the avoided crossings overlap much less than for C^{4+}/H. We may therefore expect that at least at low energies, the cross sections may be interpreted in terms of a simple sum over non-overlapping crossings. Besides, it may also be anticipated, from a comparison with the C^{3+}/H and C^{4+}/H systems, that neither translation effects nor rotational coupling will be appreciable at least for the dominant reaction channels at energies less than a few keV/amu. The Ar^{6+}/He and Ar^{8+}/He thus provide a direct test of the molecular structure calculations.

Both the Ar^{6+}/He and Ar^{8+}/He systems are treated as effective 2-electron problems with the Ar^{6+} and Ar^{8+} ionic cores represented by a suitable model potential. The method of Grice and Herschbach (1974) is then used to determine the minimum energy separation Δ_x using molecular orbitals generated by model potential techniques (Benmeuraiem et al., 1986). The results are summarized

$$Ar^{6+}/He \rightarrow Ar^{5+}(3s^2 n\ell) + He^+$$

$n\ell$	R_x (a_o)	Δ_x (eV)
4d	13.4	0.005
4p	7.4	0.52
4s	5.8	1.88
3d	3.6	6.6

$Ar^{8+}/He \rightarrow Ar^{7+}(2p^6 n\ell)/He^+$

$n\ell$	R_x (a_o)	Δ_x (eV)
5s	10.4	0.08
4f	6.42	0.52
4d	6.00	2.4
4p	4.86	4.6

For both Ar^{6+}/He and Ar^{8+}/He, we may expect, following the findings of Opradolce et al. (1983), that the LZ model should be reasonably accurate in the range from 5 eV - 1 keV/amu. It is therefore not worthwhile, at least for the present discussion, to go beyond the simple LZ model.

For Ar^{6+}/He capture occurs mainly into the 4p and 4s levels at Ar^{5+} at low energies. Capture into the 3d level becomes appreciable only at energies in excess of 200 eV/amu. The results, illustrated in Fig.11, agree remarkably well with the experimental data.

More recently, Hvelplund (1986) and McCullough et al. (1986) have extended the experimental data down to lower energies using recoil-ion techniques. Their results (as yet unpublished) confirm the good agreement between theory and experiment. Indeed the Ar^{6+}/He system provides a stringent test of the molecular model : the observed agreement between theory and experiment indicates that the set of computed values of Δ_x must be accurate to better than 20 %. Molecular structure calculations which do not have this precision may give very unreliable results.

For Ar^{8+}/He it is seen from Fig. 12 that capture at low energies takes place via the 5s, 4f and 4d states. Upwards of 50 eV/amu, capture into the 4p state also becomes appreciable, while capture into the 5s becomes negligible. The total cross-section agrees well with the experimental data of

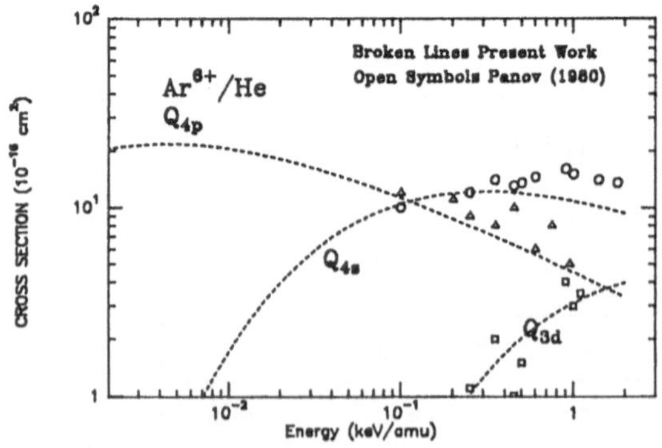

Fig. 11. Electron capture cross-sections in Ar^{6+}/He collisions

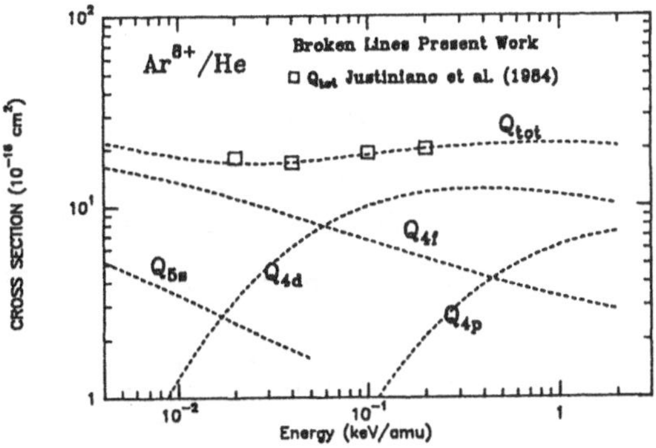

Fig. 12. Electron capture cross-sections in Ar^{8+}/He collisions

Justiniano et al. We disagree with the theoretical results of Kimura and Olson (1985) who find capture into the 4f state to be the dominant channel in the 100 eV - 1 keV/amu energy range. However, since much experimental work on Ar^{8+}/He is still in progress, it is premature to draw any definitive conclusion.

CONCLUSION

These test cases illustrate the diversity of situations which arise in applying the molecular model to the interpretation of charge transfer processes involving multiply charged ions at low energies. In spite of the formal objections to the molecular model arising from the lack of Galilean invariance of a finite adiabatic basis set, it is seen that the translation factor problem is of major concern only when we are involved with transitions occurring away from localized crossings. On the other hand, other problems arise when we have a network of crossings involving both radial and rotational couplings. The results can depend quite strongly on small energy differences between asymptotically degenerate states and very great core must be taken in the molecular calculations to ensure the dissociation energies to a very high precision. As a consequence, only a few systems in practice can be interpreted in terms of the simple LZ crossing model.

Of course, in spite of the wealth of experimental data which has come available in the last few years, there are still remarkably few systems for which a detailed comparison with theory and experiment is completely unambiguous. There is still much scope for further work.

Acknowledgements - I wish to thank Larbi Benmeuraiem, Muriel Gargaud, Liliana Opradolce and Pierre Valiron for their contribution to some of the results cited in the case studies. I am also indebted to Bob McCullough and Preben Hvelplund for communicating their experimental data prior to publication.

REFERENCES

Baliunas, S.L., and Butler, S.E., 1980, Astrophys. J. Letters, **235**, L45-48

Bates, D.R., and McCarroll, R., 1958, Proc. Roy. Soc. A, **245**, 175

Bates, D.R., and Moiseiwitch, B.L., 1954, Proc. Phys. Soc. A, **67**, 540

Benmeuraiem, L., McCarroll, R., and Opradolce, L., XI° National Conference on the Physics of Atomic and Electron Collisions, Metz, June 1986, Book of Abstracts, 40

Bienstock, S., Heil, T.G., Bottcher, C., and Dalgarno, A., 1982, Phys. Rev. A, **25**, 2850

Boyd, T.J.M., and Moiseiwitch, B.L., 1957, Proc. Phys. Soc. A, **70**, 55

Butler, S.E., and Dalgarno, A., 1980, Astrophys. J., **241**, 838

Child, M.S., 1974, Molecular Collision Theory, Academic Press, London

Ciric, D., Brazuk, A., Dijkkamp, D., De Heer, F.J., and Winter, H., 1985, J. Phys. B : Atom. Mol. Phys., **18**, 3629

Crothers, D.S.F., and Todd, N.R., 1981, J. Phys. B : Atom. Mol. Phys. **14**, 2233

Dalgarno, A., 1954, Proc.Phys. Soc. A, **67**, 1010

Delos, J.B., 1981, Rev. Mol. Phys., **53**, 287

Dijkkamp, D., Ciric, D., Vlieg, E., De Boer, A., and De Heer, F.J., 1985, J. Phys. B : Atom. Mol. Phys., **18**, 4763

Errea, L.F., Méndez, L., and Riera, A., 1984, J. Phys. B : Atom. Mol. Phys., **17**, 3271

Fritsch, W., and Lin, C.D., 1982, J. Phys. B : Atom. Mol. Phys., **15**, 1255

Fritsch, W., and Lin, C.D., 1984, J. Phys. B : Atom. Mol. Phys., **17**, 3271

Gargaud, M., Hanssen, J., McCarroll, R., and Valiron, P., 1981, J. Phys. B : Atom. Mol. Phys., **14**, 2259

Gargaud, M., and McCarroll, R., 1985, J. Phys. B : Atom. Mol. Phys. **18**, 463

Gargaud, M., et McCarroll, R., XI° Colloque sur la Physique des Collisions Atomiques et Electroniques, Metz, juin 1986, Book of Abstracts, 90

Gargaud, M., McCarroll, R., and Valiron, P., 1982, Astron. Astrophys., **106**, 197

Gargaud, M., McCarroll, R., and Valiron, P., 1986, J. Phys. B : Atom. Mol. Phys., in course of publication

Grice, R., and Herschbach, D.R., 1974, Mol. Phys., **27**, 159

Hanssen, J., Gayet, R., Harel, C., and Salin, A., 1984, J. Phys. B : Atom. Mol. Phys., **17**, L323

Hvelplund, P., 1986, private communication

Hvelplund, P., Andersen, L.H., Barany, A., Cederquist, H., Knudsen, H., and Pedersen Jok, XIV° ICPEAC, Stanford, 1986, Book of Abstracts, 474

Justiniano, E., Cocke, C.L., Gray, T.J., Dubois, R., Can, C., Waggoner, W., Schuch, R., Schmidt-Böcking, and Ingwersen, H., 1984, Phys. Rev. A, **29**, 1088

Kimura, M., and Olson, R.E., 1985, Phys. Rev. A, **31**, 489

Levy, H., and Thorson, W.R., 1969, Phys. Rev., **181**, 230

Ludde, H.J., and Dreizler, R.M., 1981, J. Phys. B : Atom. Mol. Phys. **14**, 2191

McCarroll, R., and Valiron, P., 1976, Astron. Astrophys., **53**, 83

McCarroll, R., and Valiron, P., 1979, Astron. Astrophys., **78**, 177

McCullough, R.W., Wilson, S.M., and Gilbody, H.B., 1986, J. Phys. B : Atom. Mol. Phys., in course of publication

Mittleman, M.H., 1969, Phys. Rev., **188**, 221

Opradolce, L., Valiron, P., and McCarroll, R., 1983, J. Phys. B : Atom. Mol. Phys., **16**, 2017

Panov, M.N., XI° ICPEAC, Kyoto, 1981, Invited Lectures, ed. K. Oda and K. Takayanagi, Amsterdam, North Holland, 437

Péquignot, D., 1980a, Astron. Astrophys., **81**, 356

Péquignot, D., 1980b, Astron. Astrophys., **83**, 52

Péquignot, D., Aldrovandi, S.M.V., and Stasinska, G., 1978, Astron. Astrophys. **63**, 313

Phaneuf, R.A., Alvarez, I., Meyer, F.W., and Crandall, D.H., 1982, Phys. Rev. A, **26**, 1892

Schneidermann, S.B., and Russek, A., 1969, Phys. Rev., **181**, 311
Smith, F.T., 1969, Phys. Rev., **179**, 111
Soloviev, E.A., and Vinitsky, S.I., 1985, J. Phys. B : Atom. Mol. Phys., **18**, L557
Thorson, W.R., and Delos, J.B., 1978, Phys. Rev. A, **18**, 135
Vaaben, J., and Taulbjerg, K., 1981, J. Phys. B : Atom. Mol. Phys., **14**, 1815
Wilkie, F.G., McCullough, R.W., and Gilbody, H.B., 1986, J. Phys. B : Atom. Mol. Phys., **19**, 239
Opradolce, L., Benmeuraiem, L., McCarroll, R., and Piacentini, R.D., 1987, J. Phys. B : Atom. Mol. Phys., in course of preparation

RELATIVISTIC EFFECTS IN ELECTRON CAPTURE

Benno Moiseiwitsch

Department of Applied Mathematics and Theoretical Physics
The Queen's University of Belfast
Belfast BT7 1NN, Northern Ireland

INTRODUCTION

In order to set the subject of electron capture at very high energies in historical perspective we shall first consider scattering at high non-relativistic energies.

We are concerned with electron capture from a target hydrogenic ion T, having the electron in an initial state i with wave function $\psi_i^T(\underset{\sim}{r}_T)$ and non-relativistic energy ε_T, by an incident bare projectile ion P moving in a straight line with constant velocity $\underset{\sim}{v}$ and impact parameter ρ referred to T. After capture the electron is attached to P in its final state f with wave function $\psi_f^P(\underset{\sim}{r}_P)$ and energy ε_P. Here $\underset{\sim}{r}_T$ and $\underset{\sim}{r}_P$ are the position vectors of the electron referred to the nuclei of T and P respectively.

It was originally believed that the total wave function of the electron as a function of time t could be written for slow collisions in the form

$$\Psi = a(\rho,\ t)\psi_i^T(\underset{\sim}{r}_T)e^{-i\varepsilon_T t/\hbar} + b(\rho,\ t)\psi_f^P(\underset{\sim}{r}_P)e^{-i\varepsilon_P t/\hbar} \tag{1}$$

but Bates and Stewart, in the early 1950's, found that this produces matrix elements which depend on the choice of origin, even for quite low velocities of relative motion.

Subsequently Bates and McCarroll (1958) showed that we must take into account the velocity v relative to the target nucleus T of the electron when attached to P after capture has taken place. They introduced a translation factor

$$e^{-i(\frac{1}{2}mv^2 t\ +\ m\underset{\sim}{v}\cdot\underset{\sim}{r}_T)/\hbar} \tag{2}$$

where $\underset{\sim}{v} = d\underset{\sim}{R}/dt$ and $\underset{\sim}{R} = \underset{\sim}{r}_T - \underset{\sim}{r}_P$, which was multiplied into the final atomic wave function ψ_f^P in (1). This corresponds to making a Galilean transformation from P to T.

A relativistic generalization of the above must involve the Lorentz transformation. This is achieved by representing the electron by means of the total wave function

$$\Psi = a(\rho, t)\psi_i^T(\underset{\sim}{r}_T)e^{-iE_T t/\hbar} + b(\rho, t)S\psi_f^{P'}(\underset{\sim}{r}_P')e^{-iE_P' t'/\hbar} \qquad (3)$$

referred to the inertial frame S_T of the target nucleus. Here $\psi_i^T(\underset{\sim}{r}_T)$ and $\psi_f^P(\underset{\sim}{r}_P')$ are Dirac time-independent atomic eigenvectors corresponding to eigenenergies E_T and E_P' referred to S_T and the inertial frame S_P' of the projectile ion P. Also S is an operator which transforms the eigenvector ψ_f^P referred to S_P' into an eigenvector referred to S_T.

Now the times t and t' referred to the inertial frames S_T and S_P' respectively are related by the Lorentz transformation formula

$$E_P't' = \gamma E_P't - \hbar \underset{\sim}{k}.\underset{\sim}{r}_T \qquad (4)$$

where $\underset{\sim}{k} = \gamma E_P'\underset{\sim}{v}/\hbar c^2$ is the propagation vector and $\gamma = (1 - v^2/c^2)^{-\frac{1}{2}}$.

In the non-relativistic limit

$$\gamma E_P' \to mc^2 + \varepsilon_P + \tfrac{1}{2}mv^2$$

where mc^2 is the rest mass energy of the electron, and $k \to mv/\hbar$. Thus we see that, in the non-relativistic limit, we obtain the Bates translation factor (2).

The non-relativistic Oppenheimer-Brinkman-Kramers (OBK) approximation for 1s - 1s electron capture from a target hydrogenic ion with atomic number Z_T by an incident bare projectile ion with atomic number Z_P produces the cross section

$$\sigma_{1s-1s}^{OBK} = \frac{128\pi(a_o\alpha)^2(\alpha^2 Z_T Z_P)^5}{5E(\alpha^2 Z_T^2 + q_{min}^2)^5} \qquad (5)$$

where $a_o = \hbar^2/me^2$ is the Bohr radius, $\alpha = e^2/\hbar c$ is the fine structure constant, q_{min} is the least value of the momentum change given by $q_{min}^2 \simeq E/2$ and E is the kinetic energy per atomic mass unit (amu) of the incident ion referred to the target frame. Thus as $E \to \infty$

$$\sigma_{1s-1s}^{OBK} \to \frac{2^{17}}{5}\pi(a_o\alpha)^2(\alpha^2 Z_T Z_P)^5 E^{-6} \qquad (6)$$

However the classical double scattering non-relativistic theory of Thomas (1927) leads to a capture cross section with an $E^{-11/2}$, that is v^{-11}, decay. Moreover this same fall off with energy E was discovered by Drisco (1955) using the quantal non-relativistic second Born approximation and led to the unexpected result that the second order term dominates the first order term in the Born series for electron capture. This arises from the presence of a double scattering peak in the differential cross section at the angle $\theta \sim (\sqrt{3}/2)m/M_P$ in the target frame, where M_P is the mass of the projectile ion, and corresponds to conservation of linear momentum in the first impact of the double scattering event.

Relativistic OBK Approximations

The first relativistic analysis of the OBK approximation was carried out by Mittleman (1964) who showed that the electron capture cross section must have an E^{-1} fall off with relativistic kinetic energy $E = M_P c^2(\gamma - 1)$.

A more detailed analysis at relativistic energies including numerical calculations was done by Shakeshaft (1979) and Moiseiwitsch and Stockman (1980). Moiseiwitsch and Stockman considered both capture without and with spin flip of the electron and by expanding in powers of the fine structure constant α and retaining only the lowest order terms they showed that

(1) without spin flip

$$\sigma^{OBK(1)}_{1s-1s} \simeq \frac{128}{5}\pi(a_o\alpha)^2 \frac{(\alpha^2 Z_T Z_P)^5}{\gamma - 1}\{(\alpha Z_T)^2 + q^2_{min}\}^{-5}(1 - \tfrac{1}{2}\delta^2)$$

$$\times \{1 - \tfrac{1}{2}\delta^2 + \frac{5}{18}\pi(\alpha Z_T + \alpha Z_P)\delta\} \tag{7}$$

(2) with spin flip

$$\sigma^{OBK(2)}_{1s-1s} \simeq \frac{8}{5}\pi(a_o\alpha)^2 \frac{(\alpha^2 Z_T Z_P)^5}{\gamma - 1}\{(\alpha Z_T)^2 + q^2_{min}\}^{-5}\delta^4 \tag{8}$$

where the least value of the momentum change in units of $a_o\alpha$ is given by

$$q_{min} \simeq (s_P - \gamma s_T)/(\gamma^2 - 1)^{\frac{1}{2}} \tag{9}$$

with $s_P = (1 - \alpha^2 Z_P^2)^{\frac{1}{2}}$, $s_T = (1 - \alpha^2 Z_T^2)^{\frac{1}{2}}$

and

$$\delta = \{(\gamma - 1)/(\gamma + 1)\}^{\frac{1}{2}}. \tag{10}$$

Including both capture without and with spin flip we see that

$$\sigma^{OBK}_{1s-1s} = \sigma^{OBK(1)}_{1s-1s} + \sigma^{OBK(2)}_{1s-1s} \simeq \frac{128}{5}\pi(a_o\alpha)^2 \frac{(\alpha^2 Z_T Z_P)^5}{\gamma - 1}\{(\alpha Z_T)^2 + q^2_{min}\}^{-5}$$

$$\times \{1 - \delta^2 + \frac{5}{16}\delta^4 + \frac{5}{18}\pi(\alpha Z_T + \alpha Z_P)\delta(1 - \tfrac{1}{2}\delta^2)\} \tag{11}$$

which is symmetric with respect to Z_T and Z_P since

$$(\alpha Z_T)^2 + q^2_{min} = (\alpha Z_P)^2 + q'^2_{min} \tag{12}$$

where

$$q'_{min} \simeq (s_T - \gamma s_P)/(\gamma^2 - 1)^{\frac{1}{2}} \tag{13}$$

which is the momentum change for T as projectile and P as target.

It is of interest to note that the relativistic OBK capture cross sections are modified by terms which are independent of the nuclear charges Z_T and Z_P of the ions involved in the collision. These are $(1 - \frac{1}{2}\delta^2)^2$ and $\frac{1}{16}\delta^4$ for non spin flip and spin flip collisions respectively. If Schrödinger rather than Dirac atomic wave functions are used in the analysis they become replaced by 1 and 0 respectively and must therefore be due to the interaction between the magnetic field produced by the electric current of the moving ion and the magnetic moment of the Dirac electron in the hydrogenic ion.

Relativistic Classical and Second Born Approximations

Relativistic classical double scattering theory has been used by Shakeshaft (1979) and Moiseiwitsch and Stockman (1979) who both found that the capture cross section falls off as E^{-3} and that there is a Thomas peak in the angular distribution occurring at the angle

$$\theta \sim \frac{(2\gamma + 1)^{\frac{1}{2}}}{\gamma + 1} \frac{m}{M_P} \tag{14}$$

referred to the target frame and corresponding to conservation of linear momentum. This angle approaches $(\sqrt{3}/2)m/M_P$ in the non-relativistic limit $\gamma \to 1$, as it should.

Shakeshaft (1979) showed that the relativistic classical double scattering capture cross section has the limiting form as $\gamma \to \infty$

$$\sigma_{1s-1s}^{Clas} \sim 8\pi^2(a_o\alpha)^2 \frac{(\alpha^2 Z_T Z_P)^5}{\alpha Z_T + \alpha Z_P} \frac{1}{\gamma^3} \tag{15}$$

Since the relativistic classical double scattering capture cross section decays as E^{-3} this suggests that, contrary to the non-relativistic situation, the relativistic second Born approximation will not dominate the relativistic OBK approximation which produces a capture cross section falling off as E^{-1} as $E \to \infty$.

A relativistic second Born analysis has been carried out by Humphries and Moiseiwitsch (1984). They obtained differential capture cross sections which have a double scattering Thomas peak occurring at the same angle of scattering (14) as that given by relativistic classical theory. However, unlike the non-relativistic case, another broader peak begins to develop at smaller angles of scattering as the energy increases. Indeed, as far as the contribution to the total capture cross section is concerned, the Thomas peak becomes progressively less important as $E \to \infty$ and the broad peak, which arises from the interaction between the OBK amplitude and the second Born correction amplitude, eventually dominates. At ultra high energies this broad peak has its maximum at $\theta \simeq (2/3)^{\frac{1}{2}}(\gamma^2 - 1)^{-\frac{1}{2}}m/M_P$ which approaches 0 as $\gamma \to \infty$.

Humphries and Moiseiwitsch (1984) found the following limiting formulas for the second Born approximation scattering amplitudes:

(1) without spin flip

$$g_{1s-1s}^{B2(1)} \to 8\frac{M_P}{m} a_o\alpha(\alpha^2 Z_P Z_T)^{\frac{5}{2}}(\tfrac{\gamma}{2})^{\frac{1}{2}}(\frac{1}{q^6} - \frac{1}{q^4}) \tag{16}$$

(2) with spin flip

$$g_{1s-1s}^{B2(2)} \to 8\frac{M_P}{m} a_o\alpha(\alpha^2 Z_P Z_T)^{\frac{5}{2}}(\tfrac{\gamma}{2})^{\frac{1}{2}}(q^2 - 1)^{\frac{1}{2}}(\frac{1}{q^6} - \frac{1}{q^4}) \tag{17}$$

Hence, using the cross section formula

$$\sigma = \frac{2\pi(m/M_P)^2}{\gamma^2 - 1} \int_{|q_{min}|}^{\infty} |g(q)|^2 q \, dq \tag{18}$$

they find, setting $y = q^2$, that as $\gamma \to \infty$

$$\sigma_{1s-1s}^{B2(1)} \to \frac{32\pi(a_o\alpha)^2(\alpha^2 Z_T Z_P)^5}{\gamma} \int_1^{\infty} \frac{(y-1)^2}{y^6} dy$$

$$= 16\pi(a_o\alpha)^2(\alpha^2 Z_T Z_P)^5/15\gamma \tag{19}$$

$$\sigma_{1s-1s}^{B2(2)} \to \frac{32\pi(a_o\alpha)^2(\alpha^2 Z_T Z_P)^5}{\gamma} \int_1^\infty \frac{(y-1)^3}{y^6}\, dy$$

$$= 8\pi(a_o\alpha)^2(\alpha^2 Z_T Z_P)^5/5\gamma \tag{20}$$

since $q_{min} \sim -\delta \to -1$ as $\gamma \to \infty$.

Thus the limiting form of the total capture cross section is given by

$$\sigma_{1s-1s}^{B2} = \sigma_{1s-1s}^{B2(1)} + \sigma_{1s-1s}^{B2(2)} \to 8\pi(a_o\alpha)^2(\alpha^2 Z_T Z_P)^5/3\gamma \tag{21}$$

as $\gamma \to \infty$, which is 1/3 of the limiting value of σ_{1s-1s}^{OBK}.

Eikonal Approximation

A relativistic form of the eikonal (E) approximation has been used by Eichler (1985) to obtain electron capture cross sections. It is assumed in this approximation that if $Z_T \gtrsim Z_P$ the prior form is appropriate and then the probability amplitude is given by

$$a_{fi}^E(\rho) = \frac{1}{i\hbar}\int_{-\infty}^{\infty} dt \int d\underset{\sim}{r}_T \, (S\psi_f^{P'}(\underset{\sim}{r}_P')U_f e^{-iE_P t'/\hbar})^{\dagger} V_P'(r_P')\psi_i^T(\underset{\sim}{r}_T)e^{-iE_T t/\hbar} \tag{22}$$

where

$$U_f = \exp\{\frac{i}{\hbar}\int_t^\infty V_T(r_T)\, d\tau\} \tag{23}$$

or equivalently

$$U_f = \exp\{i\alpha Z_T \frac{c}{v} \ln(vr_T + \underset{\sim}{v}.\underset{\sim}{r}_T)\} \tag{24}$$

is a phase factor introduced by the eikonal approximation to satisfy Coulomb boundary conditions, and

$$V_T(r_T) = -e^2 Z_T/r_T \tag{25}$$

$$V_P'(r_P') = -e^2 Z_P/r_P' \tag{26}$$

are the electron-nucleus interaction potentials.

By expanding in powers of the fine structure constant α and retaining the lowest order terms Eichler (1985) obtained for the 1s-1s capture cross section, summed over non spin flip and spin flip collisions:

$$\sigma_{1s-1s}^E \simeq \frac{128}{5}\pi(a_o\alpha)^2 \frac{(\alpha^2 Z_T Z_P)^5}{\gamma - 1}\{(\alpha Z_T)^2 + q_{min}^2\}^{-5} \frac{\nu_T\pi}{\sinh(\nu_T\pi)}$$

$$\times \exp\{-2\nu_T \tan^{-1}|q_{min}|/\alpha Z_T\}\, I_{1s-1s}^E \tag{27}$$

where $\nu_T = \alpha Z_T c/v$ and

$$I_{1s-1s}^E = I_0^E + I_T^E + I_P^E \tag{28}$$

with

$$I_0^E = 1 - \delta^2 + \frac{5}{16}\delta^4 + \frac{5}{4}(1 - \tfrac{1}{2}\delta^2)\eta_T q_{min} + \frac{5}{12}(\eta_T q_{min})^2 \tag{29}$$

$$I_T^E = \frac{5}{2}\pi\alpha Z_T[\tfrac{1}{9}\delta(1 - \tfrac{1}{2}\delta^2) - \tfrac{1}{14}\delta^2\eta_T] \tag{30}$$

$$I_P^E = \frac{5}{2}\pi\alpha Z_P[\tfrac{1}{9}\delta(1 - \tfrac{1}{2}\delta^2) - \tfrac{1}{9}\delta^2\eta_T - \tfrac{1}{14}\delta^2(1 - \delta^2)\eta_T + \tfrac{1}{14}\delta^3\eta_T^2] \tag{31}$$

where $\eta_T = \nu_T/\alpha Z_T$. This formula is not symmetrical with respect to Z_T and Z_P. If we let $\nu_T \rightarrow 0$ the OBK approximation to the 1s–1s capture cross section given by (11) is retrieved.

Numerical calculations using the eikonal approximation without expanding in powers of α were also carried out by Eichler (1985) and as long as Z_T is not too large good agreement was obtained with the closed analytical form (27) and (28) to (31). Moreover by allowing for shielding of the electrons in the K shell and including the contribution for capture to both K and L shells, Anholt and Eichler (1985), obtained very satisfactory accordance with experimental data on electron capture by completely stripped ions from various target atoms.

Symmetric Eikonal Approximation

It has been shown (Moiseiwitsch 1986) that the symmetric eikonal (SE) approximation for the probability amplitude for electron capture can be written

$$a_{fi}^{SE}(\rho) = \frac{1}{i\hbar}\int_{-\infty}^{\infty} dt \int d\underset{\sim}{r}_T \, (S\psi_f^{P'}(\underset{\sim}{r}_P')U_f e^{-iE_P t'/\hbar})^{\dagger} V_T(r_T) G(\hat{\underset{\sim}{r}}_T)$$
$$\times \; U_i \psi_i^T(\underset{\sim}{r}_T) e^{-iE_T t/\hbar} \tag{32}$$

where

$$G(\hat{\underset{\sim}{r}}_T) = 1 - \frac{c}{v}\,\frac{\underset{\sim}{\alpha}.(\hat{\underset{\sim}{r}}_T + \hat{\underset{\sim}{v}})}{1 + \hat{\underset{\sim}{v}}.\hat{\underset{\sim}{r}}_T} \tag{33}$$

$\underset{\sim}{\alpha}$ being the vector of the Dirac matrices, and U_f is given by (24) while

$$U_i = \exp\{-\frac{i}{\hbar}\int_{-\infty}^{t'} V_P'(r_P')d\tau'\} \tag{34}$$

or equivalently

$$U_i = \exp\{i\alpha Z_P \frac{c}{v}\ln(vr_T' - \underset{\sim}{v}.\underset{\sim}{r}_T'). \tag{35}$$

This is the post form. It has been shown (Moiseiwitsch 1986) that it is equivalent to the prior form:

$$a_{fi}^{SE}(\rho) = \frac{1}{i\hbar}\int_{-\infty}^{\infty} dt \int d\underset{\sim}{r}_T \, (S\psi_f^{P'}(\underset{\sim}{r}_P')U_f e^{-iE_P t'/\hbar})^{\dagger} V_P'(r_P')$$
$$\times \; G'(\hat{\underset{\sim}{r}}_P')U_i \psi_i^T(\underset{\sim}{r}_T) e^{-iE_T t/\hbar} \tag{36}$$

where

$$G'(\hat{\underset{\sim}{r}}'_P) = S^{-1}[1 - \frac{c}{v} \frac{\underset{\sim}{\alpha}.(\hat{\underset{\sim}{r}}'_P - \hat{\underset{\sim}{v}})}{1 - \hat{\underset{\sim}{v}}.\hat{\underset{\sim}{r}}'_P}] S^{-1} \qquad (37)$$

Assuming that αZ_T and $\alpha Z_P << 1$ it has been shown (Moiseiwitsch 1986) that for 1s-1s capture the symmetric eikonal cross section summed over non spin flip and spin flip collisions can be approximated by

$$\sigma^{SE}_{1s-1s} \simeq \frac{128}{5} \pi(a_o\alpha)^2 \frac{(\alpha^2 Z_T Z_P)^5}{\gamma - 1} \{(\alpha Z_T)^2 + q^2_{min}\}^{-5}$$

$$\times \frac{\nu_T \pi}{\sinh(\nu_T \pi)} \exp\{-2\nu_T \tan^{-1}|q_{min}|/\alpha Z_T\}$$

$$\times \frac{\nu_P \pi}{\sinh(\nu_P \pi)} \exp\{-2\nu_P \tan^{-1}|q'_{min}|/\alpha Z_P\} \, I^{SE}_{1s-1s} \qquad (38)$$

where $\nu_P = \alpha Z_P c/v$ and

$$I^{SE}_{1s-1s} = I^{SE}_0 + I^{SE}_T + I^{SE}_P \qquad (39)$$

with, correct to first order in αZ_T and αZ_P:

$$I^{SE}_0 = 1 - \delta^2 + \frac{5}{16}\delta^4 + \frac{5}{4}(1 - \tfrac{1}{2}\delta^2)(\eta_T q_{min} + \eta_P q'_{min})$$

$$+ \frac{5}{12}(\eta_T q_{min} + \eta_P q'_{min})^2 + \frac{5}{12}(3 - 2\delta^2)\eta_T q_{min}\eta_P q'_{min}$$

$$+ \frac{5}{6}\eta_T q_{min}\eta_P q'_{min}(\eta_T q_{min} + \eta_P q'_{min}) + \frac{5}{8}(\eta_T q_{min}\eta_P q'_{min})^2 \qquad (40)$$

$$I^{SE}_T = \frac{5}{2}\pi\alpha Z_T[\frac{1}{9}\delta(1 - \tfrac{1}{2}\delta^2) - \frac{1}{14}\delta^2\eta_T - \frac{1}{9}\delta^2\eta_P - \frac{1}{14}\delta^2(1 - \delta^2)\eta_P$$

$$+ \frac{1}{7}\delta^3\eta_T\eta_P + \frac{1}{14}\delta^3\eta_P^2 - \frac{1}{10}\delta^4\eta_T\eta_P^2] \qquad (41)$$

$$I^{SE}_P = \frac{5}{2}\pi\alpha Z_P[\frac{1}{9}\delta(1 - \tfrac{1}{2}\delta^2) - \frac{1}{14}\delta^2\eta_P - \frac{1}{9}\delta^2\eta_T - \frac{1}{14}\delta^2(1 - \delta^2)\eta_T$$

$$+ \frac{1}{7}\delta^3\eta_T\eta_P + \frac{1}{14}\delta^3\eta_T^2 - \frac{1}{10}\delta^4\eta_T^2\eta_P] \qquad (42)$$

where $\eta_P = \nu_P/\alpha Z_P$.

It can be readily verified that the capture cross section given by (38) with (39) to (42) is symmetric with respect to Z_T and Z_P. If we set $\nu_P = 0$ we regain the non symmetric eikonal approximation cross section given by (27) with (28) to (31).

As $\gamma \to \infty$ it is found that I^{SE}_0 and I^E_0 approach the same value 5/48 since $\eta_T \to 1$, $\eta_P \to 1$, $\delta \to 1$ and $q_{min} \to -1$, $q'_{min} \to -1$ in this limit.

However the distributions of the scattering amplitudes g^{SE} (q) and g^E (q) with momentum change q are different. Thus as $\gamma \to \infty$ it is found, assuming that αZ_T and αZ_P are sufficiently small for the eikonal pre-factors to be close to unity, that

$$g^{SE}(q) \to \frac{8M_P}{m} a_o\alpha(\alpha^2 Z_T Z_P)^{\frac{5}{2}}\left(\frac{\gamma}{2}\right)^{\frac{1}{2}} \frac{q^2 - 1}{q^5} \tag{43}$$

$$g^{E}(q) \to \frac{8M_P}{m} a_o\alpha(\alpha^2 Z_T Z_P)^{\frac{5}{2}}\left(\frac{\gamma}{2}\right)^{\frac{1}{2}} \frac{(q^2 - 1)^{\frac{1}{2}}}{q^5} \tag{44}$$

It is found that the limiting form of $g^{SE}(q)$ is the same as the amplitude given by the second Born approximation, after summation over non spin flip and spin flip collisions, whereas the limiting form of $g^{E}(q)$ is different.

It is noteworthy that the limiting SE and E capture cross sections are the same since (18) gives for $\gamma \to \infty$

$$\sigma^{SE}_{1s-1s} \to \frac{32\pi(a_o\alpha)^2(\alpha^2 Z_T Z_P)^5}{\gamma} \int_1^\infty \frac{(y-1)^2}{y^5}\, dy \tag{45}$$

$$\sigma^{E}_{1s-1s} \to \frac{32\pi(a_o\alpha)^2(\alpha^2 Z_T Z_P)^5}{\gamma} \int_1^\infty \frac{y-1}{y^5}\, dy \tag{46}$$

which are both equal to $8\pi(a_o\alpha)^2(\alpha^2 Z_T Z_P)^5/3\gamma$. Moreover this is the same as that given by the second Born approximation (21), namely 1/3 of the limiting value of σ^{OBK}_{1s-1s} as $\gamma \to \infty$.

Table 1 Cross sections (in units of 10^{-24} cm^2) for 1s-1s electron capture from target hydrogenic ions with atomic number Z_T by projectile bare ions with atomic number Z_P.

Projectile Energy E (in 10^3 MeV amu^{-1})	OBK	E	SE: Deco and Rivarola (1986)	SE: Formulas (38) to (42)
		$Z_T = 1$, $Z_P = 1$		
0.5	1.9(-13)†	6.4(-14)	3.7(-14)	3.7(-14)
1	6.6(-15)	1.7(-15)	9.6(-16)	1.0(-15)
5	2.1(-17)	2.9(-18)	2.5(-18)	2.8(-18)
10	4.4(-18)	7.5(-19)	7.5(-19)	7.8(-19)
50	3.9(-19)	1.0(-19)	1.1(-19)	1.0(-19)
100	1.7(-19)	5.1(-20)	5.0(-20)	5.0(-20)
		$Z_T = 5$, $Z_P = 1$		
0.5	5.9(-10)	1.8(-10)	-	1.0(-10)
1	2.1(-11)	5.0(-12)	2.7(-12)	2.8(-12)
5	6.8(-14)	8.6(-15)	7.0(-15)	7.7(-15)
10	1.4(-14)	2.1(-15)	2.2(-15)	2.2(-15)
50	1.3(-15)	2.9(-16)	3.0(-16)	2.9(-16)
100	5.6(-16)	1.4(-16)	1.4(-16)	1.4(-16)

† The number in brackets gives the power of ten by which the preceding number is multiplied.

OBK: Oppenheimer-Brinkman-Kramers approximation, formula (11).

E: Eikonal approximation, formula (27) with (28) to (31).

SE: Symmetric eikonal approximation.

Deco and Rivarola (1986) have recently also developed a symmetric eikonal theory of electron capture at relativistic energies. Their analysis seems to be based on a similar approach to that described here but is otherwise quite different and produces closed formulas which are rather more involved than those given in the present lecture. A comparison between the values of the 1s-1s electron capture cross sections calculated using the SE formulas (38) and (39) to (42) and calculated by Deco and Rivarola (1986) is shown in table 1. The accordance between the two symmetric eikonal approximations is seen to be satisfactory for $Z_T = 1$, $Z_P = 1$ and $Z_T = 5$, $Z_P = 1$ at all the energies displayed. However they do not agree with the non symmetric eikonal approximation of Eichler (Anholt and Eichler 1985, Eichler 1985) given by the E formulas (27) and (28) to (31) except at very high projectile energies $E \gtrsim 5000$ MeV amu^{-1}.

Table 2 Cross sections (in units of 10^{-24} cm^2) for 1s-1s electron capture from target hydrogenic ions with atomic number Z_T by projectile bare neon ions ($Z_P = 10$) having 1050 MeV amu^{-1} energy.

Z_T	OBK		E		SE
	N	F	N	F	
13	2.1(-4)†	2.0(-4)	4.0(-5)	4.0(-5)	1.7(-5)
30	1.2(-2)	1.2(-2)	1.8(-3)	1.8(-3)	6.6(-4)
47	8.7(-2)	8.8(-2)	1.1(-2)	1.1(-2)	3.6(-3)
73	3.4(-1)	4.1(-1)	4.1(-2)	4.9(-2)	1.2(-2)
92	4.2(-1)	6.3(-1)	5.7(-2)	8.5(-2)	1.6(-2)

† The number in brackets gives the power of ten by which the preceding number is multiplied.

OBK: Oppenheimer-Brinkman - Kramers, N numerical integration, F Formula (11).

E: Eikonal approximation, N numerical integration (Anholt and Eichler, 1985), F formula (27) with (28) to (31).

SE: Symmetric eikonal approximation, formula (38) with (39) to (42).

Further, in table 2, we make a similar comparison between the eikonal (E) and symmetric eikonal (SE) approximations for $Z_T = 13, 30, 47, 73, 92$ and projectile bare neon ions ($Z_P = 10$) having 1050 MeV amu^{-1} energy referred to the target frame. It can be readily seen that the effect of symmetrization is to produce a significant reduction in the 1s-1s electron capture cross section. The implications of this reduction are not obvious since the non-symmetric eikonal approximation of Eichler, with allowance for shielding of the K shell electrons and the contributions of capture into the L as well as the K shells, gives rather good agreement with the experimental data of Crawford (1979) which have been carefully analysed by Anholt (1985). The SE formula (38) with (39) to (42) has been derived assuming that αZ_T and $\alpha Z_P \ll 1$. However making this assumption turns out to be satisfactory for $Z_T < 50$ in the case of the OBK approximation, as can be seen from table 2, and yet the symmetric eikonal approximation produces a 1s-1s capture cross section which is less than half that obtained using the non symmetric eikonal approximation for $Z_T = 13, 30, 47$ and $Z_P = 10$. One might have thought that using a symmetric capture cross section formula which does not possess a post-prior discrepancy would be advantageous but this does not seem to be the state of affairs.

References

Anholt, R., 1985, Phys. Rev., A31, 3579.

Anholt, R., and Eichler, J., 1985, Phys. Rev., A31, 3505.
Bates, D. R., and McCarroll, R., 1958, Proc. Roy. Soc., A245, 175.
Crawford, H. J., 1979, Ph.D. Thesis, University of California, Lawrence Berkeley Laboratory Report No. LBL-8807.
Deco, G. R., and Rivarola, R. D., 1986, (in course of publication).
Drisco, R. M., 1955, Ph.D. Thesis, Carnegie Institute of Technology.
Eichler, J., 1985, Phys. Rev., A32, 112.
Humphries, W. J., and Moiseiwitsch, B. L., 1984, J. Phys. B: At. Mol. Phys., 17, 2655.
Mittleman, M. H., 1964, Proc. Phys. Soc., 84, 453.
Moiseiwitsch, B. L., 1986, J. Phys. B: At. Mol. Phys., 19, (in course of publication).
Moiseiwitsch, B. L., and Stockman, S. G., 1979, J. Phys. B: At. Mol. Phys., 12, L695.
Moiseiwitsch, B. L., and Stockman, S. G., 1980, J. Phys. B: At. Mol. Phys., 13, 2975.
Shakeshaft, R., 1979, Phys. Rev., A20, 779.
Thomas, L. H., 1927, Proc. Roy. Soc., A114, 561.

COMPUTER EXPERIMENTS ON ELECTRON-ION RECOMBINATION IN AN AMBIENT MEDIUM: GASES, PLASMAS, AND LIQUIDS

Wm. Lowell Morgan

Lawrence Livermore National Laboratory
and
Department of Applied Science
University of California at Davis-Livermore
Livermore, California 94550 USA

INTRODUCTION

Concern over the rate at which ions recombine with other ions or with electrons in an ambient medium followed very shortly the discovery of the electron itself in 1896. The papers published on recombination in the ninety years since are legion. This is largely due to the wide range of phenomena in which charged particle recombination is found to be important. Examples include the upper atmosphere, vapor lamps, lasers, and radiation chemistry. The importance of recombination lies in its frequently being the rate limiting step in the removal of charged particles or in the formation of important neutral species in a system. Ionic recombination processes can be grouped into the following categories: 2-body ion-ion mutual neutralization, 3-body ion-ion recombination, 2-body electron-ion dissociative recombination, 2-body electron-ion dielectronic recombination, 2-body electron-ion radiative recombination, 3-body electron-ion collisional radiative recombination, and 3-body neutral assisted electron-ion recombination. Much of the progress in thoretical understanding of recombination processes is attributable to Sir David Bates, whose first publication and at least fifty five of his more than two hundred sixty succeeding papers have dealt with recombination processes.

The subject of charged particle recombination is very diverse, the most thorough survey of the field being the book by Massey and Gilbody (1974). Of the varieties of recombination process, classical trajectory simulation has been applied to the effects of an ambient gas on ion-ion and electron-ion recombination. All of this computational research has taken place since the publication of the aforementioned review. The theory of ion-ion recombination has been very ably reviewed by Flannery (1976), who did so in a memorial volume in honor of Professor Bates' sixtieth birthday while, more recently, Bates (1985) has comprehensively reviewed theoretical, computational, and experimental research on ion-ion recombination. Aspects of electron-ion recombination in an ambient gas (neutral and electron) were reviewed by Bates (1979a,b). In this paper I discuss the research since 1979 on electron-ion recombination in an ambient molecular medium, with an emphasis on the application of computer simulation to this problem. The use of Monte Carlo computer experiments to directly simulate the physics of ion-ion recombination was pioneered by

Bates (Bates and Mendaš, 1978) who, having written many papers over the years attacking the problem theoretically, has since published ten additional papers using simulation in dealing with aspects of the problem. The computer experiment approach has helped to provide understanding of the effects of an ambient gas on ion-ion mutual neutralization, on the competition between three body recombination and mutual neutralization, and on plasma and electric field effects on ionic recombination. While not as obviously justified as in the description of heavy particle collisions, classical trajectory computer simulations can, within certain bounds, provide a quantitative description of the effects of an ambient medium - gas, plasma, or liquid - on recombination of electrons and ions.

THE RECOMBINATION PROCESS

The first application of computer simulation to recombination was to the problem of two ions recombining in a dense ambient gas. A positive and a negative ion approach each other along a coulombic potential curve (asymptotically $X^+ + Y^-$). At various internuclear separations there may be covalent curves crossing the ionic curve. At these points the ions may cross to the covalent curve and eventually separate as neutral atoms (asymptotically $X^* + Y^*$). This process is called mutual neutralization and has a probability at zero ambient gas pressure that is often accurately (Whitten, et al., 1983) given by the Landau-Zener formula. Computer simulations have shown that the mutual neutralization process is pressure dependent (Bates and Mendaš, 1978; see Bates, 1985 for further references). In the presence of an ambient gas one of the ions may collide with a neutral gas atom and lose enough energy to be bound in the potential well. This is three body recombination, which competes with mutual neutralization to a greater or lesser extent depending on the ambient gas density and the Landau-Zener probability. The theory of this clearly pressure dependent process was first discussed for high pressure transport limited recombination by Langevin (1903) and for low pressure collision rate limited recombination by Thomson (1924). In the first case the recombination rate coefficient is proportional to the inverse of the gas pressure and in the second it is directly proportional to the pressure. Recombination theory has become much more sophisticated since Thomson's time with a great amount of progress having been made recently by Bates and by Flannery (see the reviews by Bates, 1985 and by Flannery, 1976, 1982).

The ion-ion recombination process is easily and reasonably viewed classically and thus would be expected to be amenable to modeling by simulating classical trajectories on a computer. The use of classical simulation to study recombination of electrons and ions, which as a process has many similarities to ion-ion recombination, is not as obviously justified.

As mentioned in the introduction, there are several processes by which an electron and ion can recombine. Radiative and dielectronic recombination will be excluded from this discussion as they are not relevant to the temperature and density regime that interests us here. This leaves two-body dissociative and three-body electron or neutral stabilized collision radiative recombination (CRR) to be dealt with if we want to compute that rate at which electrons and ions recombine in a molecular gas. Electron stabilized CRR has been modeled, usually by solving a large number of rate equations for atomic excited state densities, extensively in the literature (see Massey and Gilbody, 1974 and Bates, 1979b for review). This is relevant to recombination in plasmas of substantially higher ion density than those under discussion here, although I will mention it again in passing in a later section.

We can picture the initial stages of neutral molecule stabilized CRR as we did three-body ion-ion recombination. The electron approaches the ion along a coulombic trajectory, losing energy as it collides with neutral gas molecules, and becomes bound to the ion. At low temperature we can neglect quantization of the Rydberg levels because the electron will become bound at only several kT below the the ionization potential (that is, in a level of very high quantum number) where the levels are closly spaced and pressure broadened. At higher temperatures the discrete nature of the lower levels becomes an important factor in the physics of the process (hence, the rate equation approach to CRR). For the high Rydberg levels we may invoke the density-of-states correspondence principle that the probability of a transition to a given final quantum state is given by the correspondingclassical probability for a transition to a state in the corresponding volume of energy space (Percival, 1982). This probability is, of course, just the quantity sampled by a classical trajectory Monte Carlo or molecular dynamics simulation.

What is the importance of two-body dissociative recombination and how do we model it? Dissociative recombination, like dielectronic recombination, involves capture of an electron into an autoionizing state followed by a stabilization process, namely, dissociation of the molecule. This is not well described classically (see Bardsley and Biondi, 1970 and Massey and Gilbody, 1974 for descriptions of and references to the quantal theory). Dissociative recombination has been modeled within the classical simulation by putting an absorbing sphere around the positive ion. The radius of this sphere is chosen to give the correct two body rate in the zero pressure limit. This model is dicussed further below.

COMPUTER EXPERIMENTS: MONTE CARLO AND MOLECULAR DYNAMICS

We use classical particle trajectories calculated by solving the equations of motion on a computer with statistical initial conditions and ion-neutral atom collision frequencies to model the dynamics of charged particle recombination. As Picasso said of art, we might say of modeling - that it is a lie that helps us see the truth. Classical trajectory models of recombination have been very helpful in this respect. Two methods of simulation have been used in the modeling of recombination: Monte Carlo and stochastic molecular dynamics.

The Monte Carlo (MC) classical trajectory method, developed in the 1940's for simulation of neutron and photon transport, was first applied to ionic recombination by Bates (Bates and Mendaš, 1978). Bates' calculations were followed in a short time by those of Bardsley and Wadehra (1980) whose method differed from that of Bates in some respects. In these simulations a succession of pairs of ions are started at a separation r_0 with velocities chosen from a Maxwellian distribution and their trajectories are computed as they interact via their mutual attraction. As they move along their trajectories the ions may collide with neutral gas molecules with collision frequencies sampled from an appropriate statistical distribution. Recombination is said to have occurred if the total energy of ions $E \ll -kT$. If the ions separate again past r_0 (Bardsley and Wadehra, 1980) or, equivalently, if a so-called partial parting occurs (Bates and Mendaš, 1978) the trajectory is terminated. The ratio of the number of recombinations to the total number of trajectories simulated is the recombination probability, from which a rate coefficient can be computed.

The molecular dynamics (MD) method, which was developed in the late 1950's for simulating the equilibrium statistical mechanics of liquids, was first applied, using a formulation suitable for non-equilibrium systems, to recombination processes by Morgan, et al. (1982). These simulations involve solving the equations of motion of N ions in a cubic cell with periodic boundary conditions and dimensions chosen to yield the desired ion number density. Nearest images of the particles are used in the calculation of interionic forces. The equations of motion are integrated using the following algorithm (Scofield, 1973):

$$r(t+dt) = r(t) + v(t)dt + [4a(t)-a(t-dt)]dt^2/6$$

$$v(t+dt) = v(t) + [2a(t+dt)+5a(t)-a(t-dt)]dt/6$$

This simulation, unlike traditional MD methods, is known as stochastic molecular dynamics because there is a random aspect to the dynamics. This arises because we allow the ions, whose trajectories we are computing, to collide with an appropriate frequency with neutral gas atoms. This frequency is the Langevin constant collision frquency for heavy ion collisions with non-polar neutral atoms or molecules. For electrons colliding with neutrals we use the available collision cross sections for elastic, inelastic, and superelastic collisions along with the null collision method (Lin and Bardsley, 1978 and references contained therein), which allows us to apply the constant mean free time (i.e., constant collision frequency) algorithm to any arbitrary velocity dependent cross section. Clearly we need only follow the ions in the simulation and not the neutral molecules. This stochastic MD method then simulates a constant (T,V,N) canonical ensemble where the neutral particles function as a thermal reservoir maintaining the system at a constant temperature. The choice of N (100 ions in our calculations) is determined by what we consider to be the allowable temperature fluctuation (proportional to $N^{1/2}$) in the system. In addition, as ions disappear from the volume due to recombination, new ion pairs may be created in the unit cell such that the ion density is held constant.

There are three approaches available for treating recombination in the MD simulations: (i) we can say that the ions have recombined if $r<r_0$ (the absorbing sphere model); (ii) we can declare recombination when $E<<kT$ within some r_0; and (iii) we can perform the two body MC simulation for any pair of ions that approach to within separation r_0. These options allow us a great amount of flexibility under a wide range of conditions. The most important advantages of the MD simulation over the two body MC calculation are the ability to include the effects of collective (plasma) interactions and the effects of external fields on recombination processes. In addition, we do not need to use transport theory and its associated approximations to provide the boundary conditions at the perimeter of the simulation region as is necessary with the MC method. Finally, as we shall see later, we can simulate non-homogeneous processes, such as geminate recombination, using the many body, time-dependent molecular dynamics technique.

RECOMBINATION OF ELECTRONS AND IONS AT MODERATE PRESSURES

Bates (Bates and Khare, 1965; Bates, et al., 1971) had considered the effects of an ambient gas on electron recombination well in advance of truely definitive experimental measurements of such effects. Recently such effects were observed in the important experiments of Warman, et al. (1979), Sennhauser, et al. (1980), and Armstrong, et al. (1982). They have measured electron recombination rate coefficients as functions of pressure and temperature in a variety of molecular gases. The recombination rates were obtained by microwave absorption measurement of the decay of conductivity after pulse ionization of the gas. Their measured values of the recombination rate coefficient, $\alpha(cm^3\ s^{-1})$, for electrons in CO_2, H_2O, and NH_3 are shown in Fig. 1.The rate coefficient in the zero pressure limit is just that for dissociative recombination, α_2^0. At higher pressure, yet below that of the maximum rate coefficient, α increases linearly with pressure.

Warman, et al. (1979) plot $\alpha-\alpha_2^0$ verus p and find a linear pressure dependence indicating a three-body Thomson type of process. At very high pressure the rate of electron-ion recombination is simply limited by the rate of transport of electrons toward ions. This is the Langevin-Harper-Debye diffusion limit and is expressed in terms of the mobility, μ, and the dielectric constant, ϵ, as (Bates, 1975; 1983),

$$\alpha_L = 4\pi e\mu/\epsilon \qquad (1)$$

It is the intermediate pressure region that is most difficult to describe quantitatively and has been the subject of most of the theoretical and computational research.

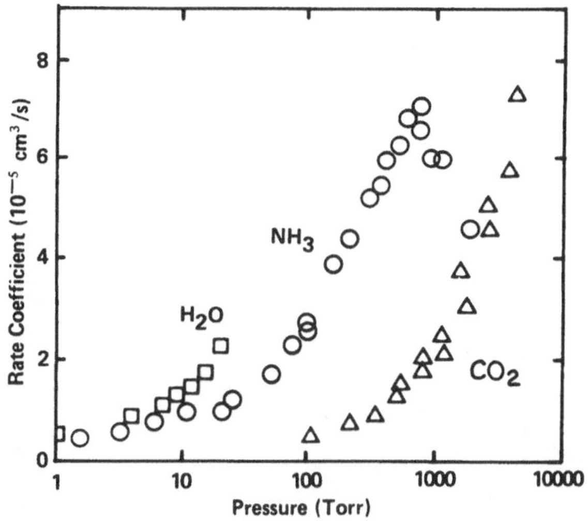

Fig. 1. Measured recombination rate coefficients for electrons in molecular gas (Warman, et al.,1979; Armstrong, et al., 1982).

Bates has attacked the problem with classical theory (Bates, 1980a) followed by a quantal approach (Bates, 1981a). In the classical theory he considered only three body recombination obtaining very reasonable results but noted that the predicted temperature dependence was incorrect and that the quantal theory needed to correct this problem would yield much lower and generally unsatisfactory rate coefficients. With the quantal theory Bates considered dissociative recombination in addition to the three body process. In doing this he introduced a process that he called *collisional dissociative recombination* (CDR) whereby electrons in Rydberg levels may be take part in a radiationless transition leading to molecular dissociation. The important ingredients of the theory are $P(\ell)$, the probability of dissociative recombination during a single passage of the Rydberg electron of angular momentum ℓ through the inner region of the system, and the so-called coupling, the rms change in ℓ per collision. Bates (1981a) did calculations assuming a range of average values for $P(\ell)$ and weak and strong coupling limits. With this model CDR becomes the dominate recombination mechanism and qualitative agreement with the measured data is achieved. In view of the complexity of the chain of processes involved and the approximations made, Bates considers the agreement to be satisfactory and states that the measurements no longer pose a problem. Bates (1982) later raised another serious question in comparing ion-ion and electron-ion recombination.

Bates' Observation

In their Monte Carlo simulations of ion-ion recombination Bardsley and Wadehra (1980), as I have mentioned above, computed a recombination probability, w, for two ions approaching one another from afar. This probability multiplies the flux of ions, obtained from a Fick's law mobility-diffusion transport equation, passing through a separation radius r_0 to obtain a rate coefficient. The final expression for the rate coefficient is

$$\alpha = [X/\alpha_T + (1-X)/\alpha_L]^{-1} \quad (2)$$

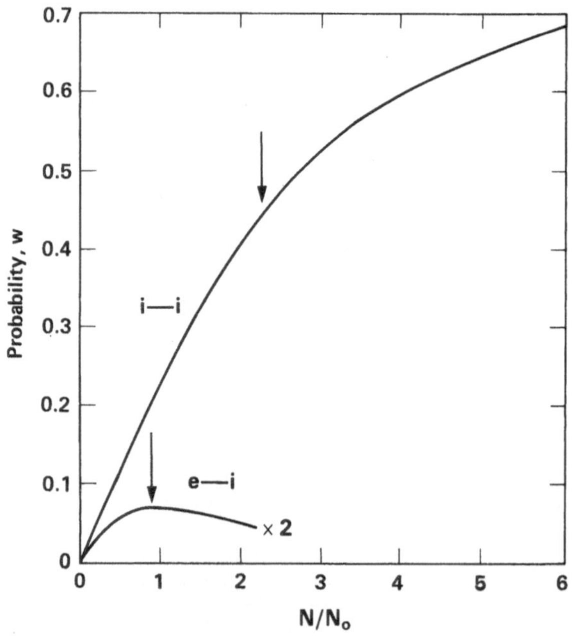

Fig. 2. Ion-ion ($O_4^+ + O_4^-$) and electron-ion (in NH_3) recombination probabilities. Arrow indicate density of maximum in rate coefficient.

where $X = \exp(-e^2/r_0kT)$ and

$$\alpha_T = \pi r_0^2 (8kT/\pi M)^{1/2} w = \alpha_0 w \qquad (3)$$

α_T and α_L are the Thomson (low pressure) and the Langevin (high pressure) rate coefficients respectively.

Bates (1982) has used this formulation and the measured rate coefficients for electrons in NH_3 (Armstrong, et al., 1982) to obtain recombination probabilities as a function of ambient gas density.

$$w = \alpha/\alpha_0(Y-Z) \qquad (4)$$

where $Y=1/X$ and $Z=\alpha(Y-1)/\alpha_L$. He then compared this probability function to that obtained (Bates and Mendaš, 1982) from Monte Carlo simulations of O_4^+-O_4^- recombination in oxygen. His results are shown in Fig. 2. Two observations stand out. First, the w values for electron-ion recombination are very much smaller than those for ion-ion recombination. Second, w(N) peaks at the same density at which the rate coefficient peaks for electrons but is still increasing with density for ions. Bates notes that for recombination of positive and negative ions the cancelation between Z and Y in the denominator of Eqn. (4) becomes pronounced causing w to continue rising beyond the maximum in α. What, then, is so very different about electrons recombining with ions in an ambient gas? Bates (1982) has suggested that the early fall of w(N) may be a result of the weakening due to cluster formation of the collisional coupling between dissociative states and other states.

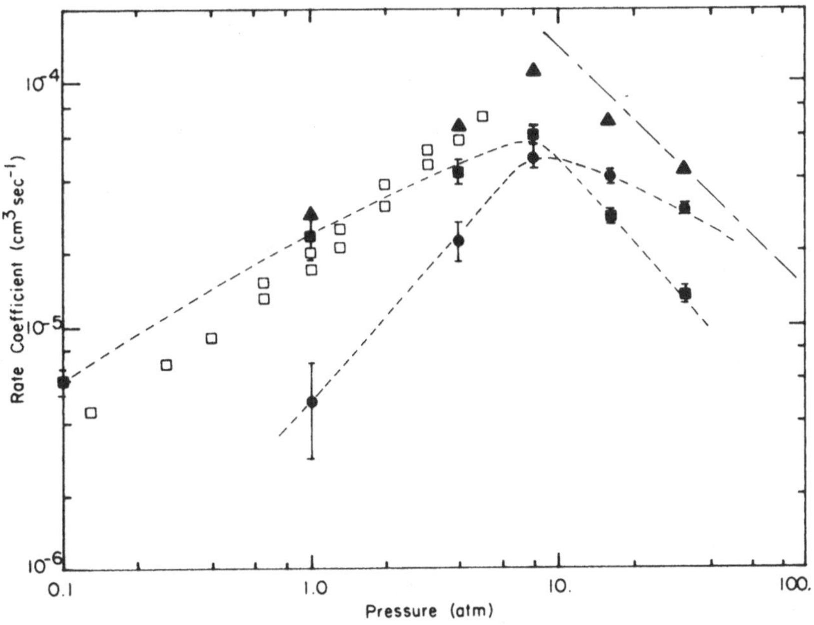

Fig. 3. Recombination in CO_2. Points ■ and ● are the 2-body and 3-body contributions to the total ▲ calculated coefficient. The points □ are the measured values of Armstrong, et al. (1982)

Results of Computer Simulations

Monte Carlo simulations have been performed for recombination in CO_2, NH_3, and H_2O (Morgan and Bardsley, 1983 and Morgan, 1984). Results for carbon dioxide and for ammonia are shown in Figs. 3 and 4 along with the measured values that we have already seen in Fig. 1. The details of the cross sectional data used in these calculations are given in Morgan and Bardsley (1983).

The importance of inelastic collisions is apparent from the CO_2 calculation without and with vibrational excitation (0.083 eV threshold energy). Rotational excitation is the most important collisional process in NH_3 which has very large cross sections, of order 200 a_0^2 at thermal energies. We see that the calculated total recombination rate coefficients are reasonably accurate. The recombination model indicates that dissociative recombination is enhanced by electron collisions with the ambient gas molecules and is the dominate form of recombination at low and intermediate pressures. This is, of course, just the collisional dissociative recombination process proposed by Bates (1981a). The sequence of events in the simulation leading to CDR is inelastic electron-molecule collisions leaving the electrons in high bound orbits (Rydberg levels) with small angular momentum ℓ followed by

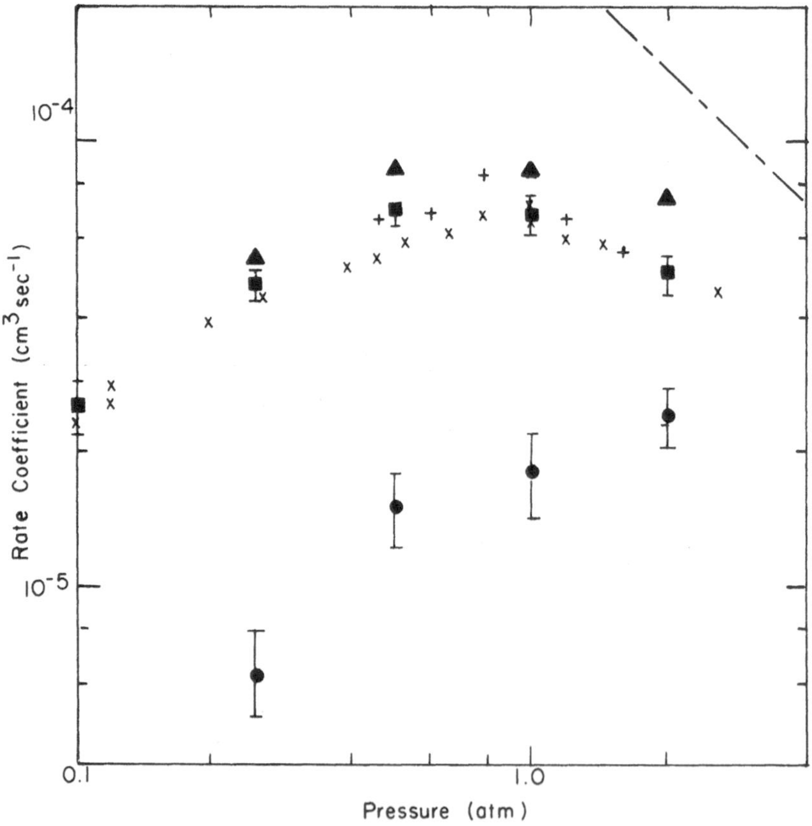

Fig. 4. Recombination in NH_3. Points ■ and ● are the 2-body and three body contributions to the total rate coefficient ▲. Points X and + are the measurements of Armstrong, et al. (1982) and Sennhauser and Armstrong (1978) respectively.

dissociative recombination. The absorbing sphere used in the simulation preferentially selects electrons in low ℓ orbits for dissociative recombination. We can only speculate on whether or not this $\alpha_2(N)$ is correct because there are no experiments that have separated the dissociatve and the three body channels. We would expect there to some enhancement because the theoretical dissociative recombination rate coefficient has a $T^{-3/2}$ temperature dependence (Bardsley and Biondi, 1970; Massey and Gilbody, 1974). It would be interesting to know how well this crude absorbing sphere model of dissociative recombination mimics reality.

We have now seen that classical trajectory simulations yield reasonable results for electron-ion recombination in an ambient gas as they had for ion-ion recombination. We have also examined Bates' observation indicating some fundamental difference in the physics of the two processes. Since the calculations appear to be a good representation of the physical process they must contain in some way the answers to the questions posed by Bates (1982). This is discussed next.

A Proposal

In his original theory of ionic recombination Thomson (1924) assumed that recombination would occur upon a collision when the two ions were separated by a distance $r<r_T$. This happens because an ion colliding with a neutral gas molecule would lose most of its kinetic energy (the fractional energy loss is $4M_iM_n/(M_i+M_n)^2$ for $v_i>>v_n$) and, hence, emerge from the collision with total energy $E<<kT$; that is, it would be bound to the other ion. In the classical equivalent picture of electron recombination it is *inelastic*, rather than elastic, collisions that are responsible for the energy loss leading to recombination. Inelastic collisions are generally unimportant in ion-ion recombination (Bates, 1981b). The electron-molecule inelastic cross sections generally have a threshold energy and a complicated energy dependence. The electron mean free path for inelastic collisions having threshold energy u_{th} is given by

$$\lambda_{inel} = [N \int_{u_{th}}^{\infty} \sigma_{inel}(u)\ f(u)\ u^{1/2}du]^{-1} \qquad (5)$$

Consequently the rate of energy loss by electrons within the Thomson radius will depend strongly upon the shape of the electron energy distribution function, f(u). In a Monte Carlo simulation of recombination in NH_3 I have sampled the energies of electrons passing through a shell of radius $r_T/2$ surrounding a positive ion and have obtained the f(u) shown in Fig. 5.This behavior will be typical of electrons in molecular gases. The low density limit is obviously just a Maxwell-Boltzmann distribution modified by the kinetic energy that electrons gain in coming from $r=\infty$ to $r=r_T$. In the high density limit the distribution is a Maxwellian at the gas temperature. As the electron mean energy approaches the gas mean energy with increasing gas density the effect of inelastic collisions diminishes due to detailed balance in addition to the reduction in the fraction of electrons above the excitation threshold energy.

We can estimate the recombination probability using an expression derived by Loeb (1955) for the probability of a single collision within radius r_T of an oppositely charged ion (assuming rectilinear trajectories). For a mean free path λ and letting $x=2r_T/\lambda$ the collision probability w is

$$w = 1 - (2/x^2)[1 - (x+1)e^{-x}] \qquad (6)$$

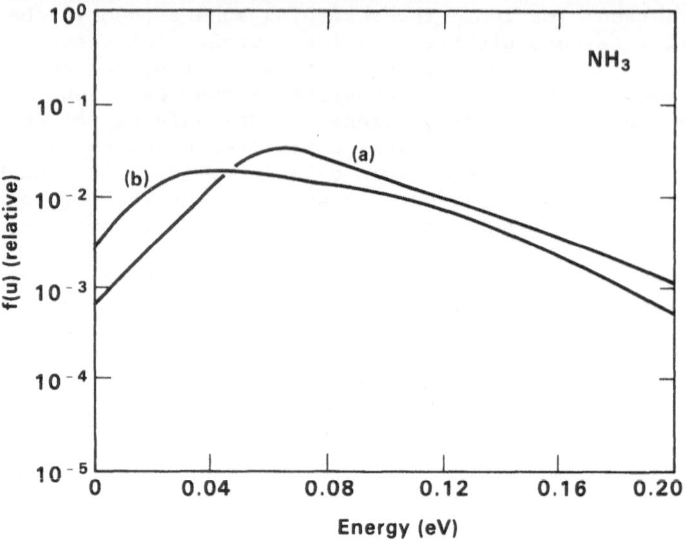

Fig. 5. Energy distribution of electrons at radius $r_T/2$ from ion in NH_3 at (a) 0.5 atm and (b) 2 atm.

For ions in CO_2 at 1 atm pressure and 300K the mean free path based on the Langevin ion-induced-dipole collision frequency is 440 a_0. The Thomson radius is 718 a_0. Eqn. (6) then gives w=0.83. Hence if all ions recombined on their first encounter within r_T the recombination probability would be 83%. This probability is of the same order as that shown in Fig. (2) and, as Bates (1982) points out, is to be expected for ion-ion recombination.

Using the momentum transfer cross section, which is proportional to $E^{-1/2}$ for energies less than about 0.2 eV, for electrons in CO_2 (Hake and Phelps, 1967) gives a mean free path of 470 a_0 for thermal electrons in atmospheric pressure CO_2. This is of little consequence, however, because the energy loss per collision is negligible. The vibrational excitation threshold, on the other hand, is at 0.083 eV ≃ 3kT and has a cross section of approximately 11 a_0^2 (Hake and Phelps, 1967). This gives a mean free path of 2.4×10^4 a_0 for electrons in the energy range 3kT<E<6kT. The inelastic collision probability is then 0.035, which is consistent with the value of 0.01 that is obtained using Eqn. (4). Since in the low density limit the average electron coming from infinity to r_T will have kinetic energy 3kT this may be representative of the recombination probabilties to be expected in electron-ion recombination, noting that an electron will lose 3kT of energy in a single inelastic collision. This probability is of the same order as Bates (1982) has deduced for electron recombination in NH_3.

Bates (1982) points out that the electron recombination probability, in contrast to that for ions, peaks at the same density as the rate coefficient. By comparing the NH_3 rate coefficients of Fig. (4) and the electron distributions functions of Fig. (5) we might surmise that α peaks at the same density as w and that w peaks where it does because of the effects of density on f(u) and, hence, λ_{inel}. Observe that for NH_3 at 2 atm f(u) is still far from being a thermal distribution at the gas temperature. At large enough densities in all gases f(u) will approach thermal. At these densities we would expect recombination to be determined by the drift rate and to be given by the Langevin formula Eqn. (1).

A successful prediction: recombination in a discharge

The computer simulations demonstrate the importance of energy exchange collisions in the recombination process. The MC and MD simulations appear to give, subject to the availability of accurate cross sectional data, reasonable quantitative predictions for the electron-ion recombination rate coefficient. There are exceptions, however, as we will see below in the discussion of recombination in methane. An interest in recombination of electrons in high pressure carbon dioxide lasers and in electron beam controlled discharge switches led to MD calclulations of plasma and electric field effects on α in a weakly ionized discharge. These calculations (Morgan, 1984b) demonstrated some non-equilibrium and plasma screening effects on α, which were eventually overwhelmed by the CRR rate at high electron density, but, more importantly, showed the effects of the electric field in a discharge on the recombination rate. Those results are plotted in Fig. (6).

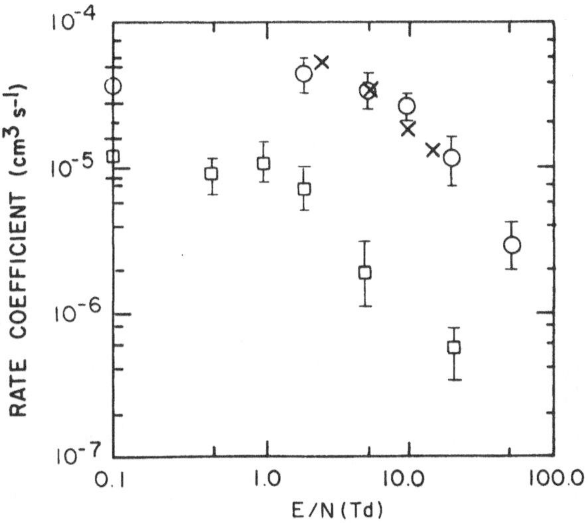

Fig. 6. Recombination in a discharge. Molecular dynamics simulations for electrons in CO_2 (○) and CH_4 (□). Points X are measurements of Littlewood, et al. (1983).

Littlewood, et al. (1983) have measured electron recombination rates in $CO_2/N_2/He$ mixtures using a pulsed e-beam sustained discharge. Measurement of the discharge current plus knowledge of the e-beam ionization rate and the electron drift velocity yields the recombination rate. The CO_2 measurements of Littlewood, et al. (1983) are also plotted in Fig. (6) where we see excellent agreement. In the process of modeling an high pressure switch in methane Kline (1982) has estimated the recombination rate coefficient to be about $2x10^{-6}$ cm^3/s for E/N values of a few Td. This estimate agrees with the MD calculation of for CH_4 in an electric field, as shown in Fig. (6). Unfortunately, as we will see next, these methane results presented are dubious in light of recent recombination measurements by Nakamura, et al. (1983).

Nakamura, Shinsaka, and Hatano [NSH] (1983) have recently measured, using pulse radiolysis techniques, the electron mobility and recombination rate in gaseous, liquid, and solid methane! We see in Fig. (7) the rate coefficients measured by NSH.They also plot the ratio α/α_L as a function of density and show that in the gas phase the recombination rate ranges from about 15% of the diffusion limited rate at the lowest densities (about 10 atm) to a maximum of about 70% of α_L at the critical point (about 300 atm). I will discuss the liquid phase results shortly.

We see that the results of computer simulation fall far below the measured rates (and, of course, the Langevin limiting rate) for recombination in gaseous CH_4 but that they are in fair agreement with the measurement for densities above about 10^{21} cm^{-3}. The rate is expected to peak at very large N because the inelastic threshold is at about 4kT. There are some possible theoretical causes for uncertainty at these high gas densities (as well as, of course, possible experimental difficulties). Examination of the static structure factor S(k=0,N), which is proportional to the isothermal compressibility, shows that structure effects (significant pair correlations) set in at densities of several times 10^{21} cm^{-3} (Christophorou and McCorkle, 1976).

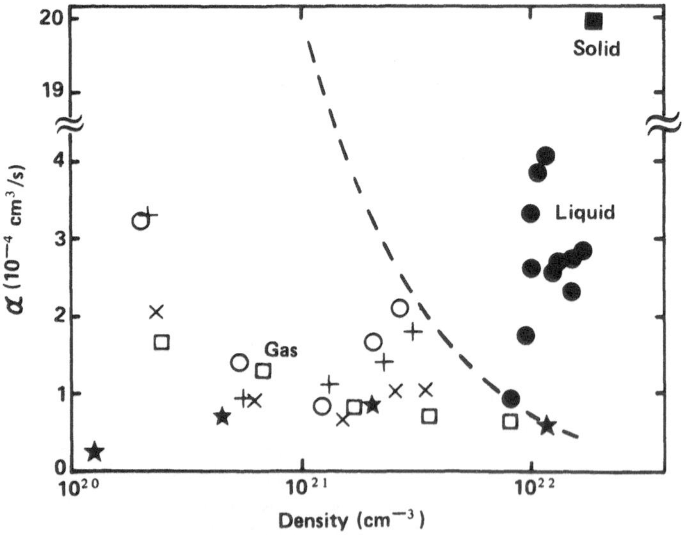

Fig. 7. Rate coefficients for recombination in methane measured by Nakamura, et al. (1983). The gas phase rates were measured at four temperatures: □ 193K; X 222K; + 254K; and O 295K. The points ★ are the MD calculated rates and the dashed line ----- is the Langevin diffusion limited rate coefficient of Eqn. (1).

Density effects are sometimes inluded in mobility calculations in an approximate manner by using a Born approximation result for the momentum transfer cross section (Atrazhev and Yakubov, 1981),

$$\sigma_m(N) = \sigma_m(N=0)S(k=0,N)$$

As Braglia and Dallacasa (1982) have discussed, however, this approach is only very approximate and uncertainties in cross sectional data and high density effects conspire to make theoretical electron drift velocities qualitatively as well as quantitatively incorrect for densities as low as several times 10^{20} cm^{-3}. Part of the difficulty is due to the very large uncertainty in the magnitude and energy dependence of the elastic cross section for slow electrons in CH_4, especially near the Ramsauer minimum (Ferch, et al., 1985). The uncertainty persits despite numerous cross section measurements and computations dating back to 1925! This uncertainty along with the difficulty in describing the effects of density on the elastic and inelastic cross sections (see Davis, et al., 1971 for discussion of coherent and incoherent contributions to scattering in dense media) can have a large effect on the recombination rate calculations because the inelastic cross section in methane is several times the elastic cross section in the vicinity of the Ramsauer minimum (Pitchford, et al., 1981).

Recombination in liquid methane

The pulse radiolysis measurements of NSH involve placing the liquid in a conductance cell and irradiating it with a burst of x-rays. If a voltage is applied across the cell and the time dependence of the current measured, the electron mobility and homogeneous recombination rate coefficent can be extracted from the decay of the conductivity signal. Each high energy photon passing through the liquid produces a track of ionized molecules. The electron in each electron-ion pair is initially speeding away from the parent ion but, in time, loses enough energy in collisions with the surrounding molecules to be thermalized. Upon thermalization the electron is likely to recombine with a nearby ion. Recombination with the parent ion is termed *geminate* whereas recombination with a different ion is called *homogeneous*. This process was orgininally described theoretically by Onsager (1938) who applied the Smoluchowski equation to the Brownian motion of a particle under the action of a Coulomb field plus a collecting field in the presence of sinks at the origin and at infinity. This is the standard model used in the analysis of pulse radiolysis experiments.

Electrons in liquid methane are quasifree resulting in very high mobility, of the same order as in liquid Ar, Kr, and Xe, which are the standard subjects of research on electron transport in liquids. We would expect the recombination rate coefficient to be diffusion limited because of the density of the medium and to be rather large because of the large mobility. We see in Fig. (8) that NSH have measured very large rate coefficients, but there is a surprise. Except near the critical point, they find that α is substantially smaller than α_L. Their measured rate coefficients and the diffusion limited values are shown in Fig. (8).

The molecular dynamics method has been used (Morgan, 1986) to simulate electron transport and recombination in liquid CH_4. The modeling of the process is similar to what was done in previous work (Morgan, et al., 1982; Morgan, 1984b) except for the details of the ionization of the liquid and the distinction between geminate and homogeneous recombination. At the start of the simulation or after a recombination, the positive ions having been randomly placed in the simulation volume, an electron is placed at radius e^2/I_p from its companion ion and given an initial velocity radially outward. Its initial kinetic energy is I_p+T_s, where T_s is chosen from a standard distribution $f(T_s)$ of secondary electron energies (Green and Sawada, 1972). The electrons and ions are labelled so that the two types of recombination can be tallied separately.

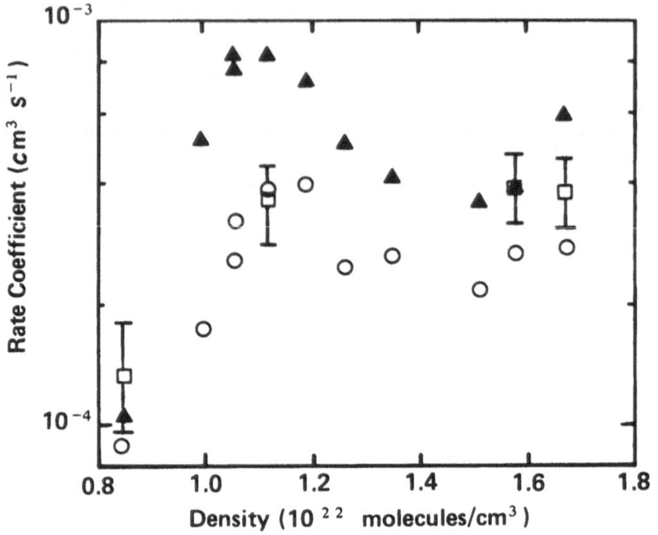

Fig. 8. Recombination in liquid methane. Points ○ are the rates measured by Nakamura, et al. (1983); □ are MD simulations of Morgan (1986); and ▲ are the Langevin rate coefficients from Eqn. (1).

The usual model cross section for methane vibrational excitation (Pitchford, et al., 1981) was used in the simulation. Due to the great density of a liquid the individual atomic potentials responsible for elastic scattering are smoothed out so that the elastic scattering of quasifree electrons is due to deformation potentials or phonons as is scattering in semiconductors. There are several models for such scattering in liquids (Berlin, et al., 1978; Basak and Cohen, 1979; Ascarelli, 1986). For the sake of accuracy, a mean free path for phonon scattering obtained from the measured mobility was used in the MD simulations. The results of these calculations are also plotted in Fig. (8). We see that the calculated rate coefficients agree well with the measured values, much better than we saw in the gas phase comparison. The uncertainties in the elastic scattering model, in the inelastic cross section, and possibly in the electron effective mass (recall that we are treating electrons in the conduction band of a condensed medium) are large enough that the calculated α is more accurate than we may rightfully expect it to be. The results of these calculations are consistent with the rate of collisional energy loss within the reaction radius rather than diffusion being the rate limiting step in recombination. This, of course, leads to $\alpha<\alpha_L$. Finally, the computed electron thermalization time of 6.3ps, obtained by sampling the speeds of electrons as they slow down, agrees well with another estimate (Warman, 1981).

One final comparison can be made between simulation and measurement for electrons in liquid methane. The probability of geminate recombination P_{gr} has been measured by Freeman and collaborators and is plotted below in Fig. (9) (Gee and Freeman, 1986) with the values obtained from the MD simulations (Morgan, 1986).

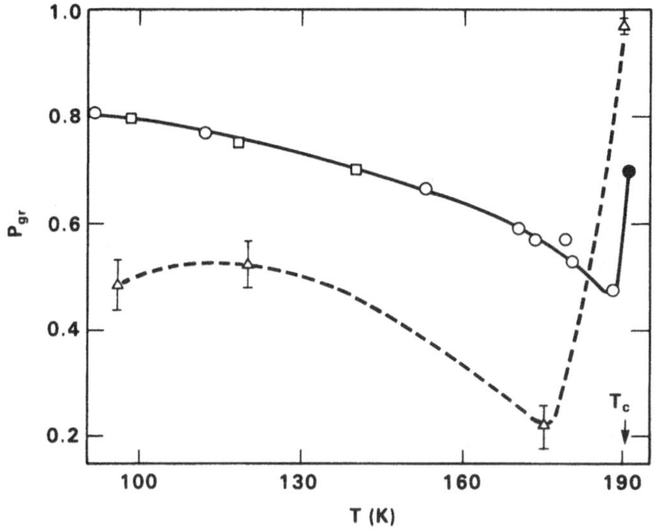

Fig. 9. Probability of geminate recombination in liquid methane. Points □, ○, and ● are mesurements from Gee and Freeman (1986) and △ are MD calculations of Morgan (1986).

Although the abscissa in the graph is temperature it represents density variation because the liquid is under its vapor pressure. The temperature and density (in units of 10^{22} cm^{-3}) correspondences for the four computed points are (96K, 1.675), (120K, 1.545), (175K, 1.122), and (190K, 0.851). The critical point is at T=190K. The qualitative resemblence of the two curves is apparent. We might expect P_{gr} to be more sensitive to the details of the ionization model and the magnitude and shape of the inelastic cross section than the homogeneous recombination rate coefficient. As for P_{gr} at the critical point, the physics of a liquid at the critical point is very complex (Ziman, 1979) and we might expect this model of electron transport and recombination to be too crude to expect much predictive accuaracy. We see, however, that computer simulation of this complex process, which we might think of as an *ab initio* approach, does not give unreasonable results.

FINAL REMARKS

The path followed by Bates (1980a, 1981a) in his treatments of electron recombination, first the classical and then the quantal theory and the introduction of CDR, the dilema posed by Bates (1982), and its proposed resolution demonstrate that electron-ion recombination in the presence of a high density ambient molecular gas or liquid is a much more complex theoretical problem that ion-ion recombination. On the other hand, the similarity of the α(N) curves for the two processes provides hope that, given the proper parameters, one may still be able to construct a "universal curve" for α(N) as Bates (1980b) has done for ion-ion recombination. In addition to further theoretical or computational work, noting that I have not discussed temperature dependence and quantum effects or the effects of clustering, there is need for further measurements in the intermediate and high density regions.

ACKNOWLEDGEMENTS

I am pleased to acknowledge the contributions of my collaborators, Dr. J. Norman Bardsley and Dr. Barbara L. Whitten, to our research on recombination processes. I wish also to thank Professor Sir David R. Bates for discussions from which we profited greatly and which, sometimes, kept us from going astray in our research. This work was performed in part under the auspices of the U.S. Department of Energy at the Lawrence Livermore National Laboratory under Contract No. W-7405-Eng-48.

REFERENCES

Armstrong, D. A., Sennhauser, E. S., Warman, J. M., and Sowada, U., 1982, The electron-ion recombination coefficient in CO_2 and NH_3. Deviations from a linear density dependence at elevated pressures, Chem. Phys. Lett., 86:281.

Ascarelli, G., 1986, Calculation of the mobility of electrons injected in liquid methane, Phys. Rev. B, 34:7329.

Atrazhev, V. M., and Yakubov, I. T., 1981, Electron mobility in liquids and dense gases, High Temperature (USSR), 18:969.

Bardsley, J. N., and Biondi, M. A., 1970, Dissociative recombination, in: "Advances in Atomic and Molecular Physics," D. R. Bates and I. Esterman, eds., Academic Press, New York.

Bardsley, J. N., and Wadehra, J. M., 1980, Monte Carlo simulation of three-body ion-ion recombination, Chem. Phys. Lett., 72:477.

Basak, S. and Cohen, M., 1979, Deformation-potential theory for the mobility of electrons in liquid argon, Phys. Rev. B, 20:3404.

Bates, D. R., 1975, Ionic recombination in a high density ambient gas, J. Phys. B, 8:2722.

Bates, D. R., 1979a, Aspects of recombination, in: "Advances in Atomic and Molecular Physics," D. R. Bates and B. Bederson, eds., Academic Press, New York.

Bates, D. R., 1979b, Electron-ion recombination in ambient electron and neutral gases, in: "The Physics of Ionized Gases," R. K. Janev, ed., Institute of Physics, Belgrade.

Bates, D. R., 1980a, Classical theory of electron-ion recombination in an ambient gas, J. Phys. B, 13:2587.

Bates, D. R., 1980b, Universal curve for ter-molecular ionic recombination coefficients, Chem. Phys. Lett., 75:409.

Bates, D. R., 1981a, Electron-ion recombination in an ambient molecular gas, J. Phys. B, 14:3525.

Bates, D. R., 1981b, Effect of inelastic collisions on rate of termolecular ionic recombination, J. Phys. B, 14:2853.

Bates, D. R., 1982, Electron-ion collisional dissociative recombination at high ambient ammonia densities, Chem. Phys. Lett., 89:294.

Bates, D. R., 1983, Termolecular ionic recombination at high ambient gas density, J. Phys. B, 16:L295.

Bates, D. R., 1985, Ion-ion recombination in an ambient gas, in: "Advances in Atomic and Molecular Physics," D. R. Bates and B. Bederson, eds., Academic Press, New York.

Bates, D. R., and Khare, S. P., 1965, Recombination of positive ions and electrons in a dense neutral gas, Proc. Phys. Soc., 85:231.

Bates, D. R., Malaviya, V., and Young, N. A., 1971, Electron-ion recombination in a dense molecular gas, Proc. Roy. Soc. Lond. A, 320:437.

Bates, D. R., and Mendaš, I., 1978, Ionic recombination in an ambient gas II. Computer experiment with specific allowance for binary recombination, Proc. Roy. Soc. Lond. A, 359:287.

Bates, D. R., and Mendaš, I., 1982, Rate coefficients for ter-molecular ionic recombination, Chem. Phys. Lett., 88:528.

Berlin, Y. A., Nyikos, L., and Schiller, R., 1978, Mobility of localized and quasifree excess electrons in liquid hydrocarbons, J. Chem. Phys., 69:2401.

Braglia, G. L., and Dallacasa, V., 1982, Theory of electron mobility in dense gases, Phys. Rev. A, 26:902.

Christophorou, L. G., and McCorkle, D. L., 1976, Experimental evidence for the existence of a Ramsauer-Townsend minimum in liquid CH_4 and liquid Ar (Kr and Xe), Chem. Phys. Lett., 42:533.

Davis, H. T., Schmidt, L. D., and Minday, R. M., 1971, Kinetic theory of excess electrons in polyatomic gases, liquids, and solids, Phys. Rev. A, 3:1027.

Ferch, J., Granitza, B., and Raith, W., 1985, The Ramsauer minimum of methane, J. Phys. B, 18:L445.

Flannery, M. R., 1976, Ionic recombination, in: "Atomic Processes and Applications," P. G. Burke and B. L. Moiseiwitsch, eds., North-Holland, Amsterdam.

Flannery, M. R., 1982, Ion-ion recombination in high pressure plasmas, in: "Applied Atomic Collision Physics V.3: Gas Lasers," E. W. McDaniel and W. L. Nighan, eds., Academic Press, New York.

Frost, L. S., and Phelps, A. V., 1962, Rotational excitation and momentum transfer cross sections for electrons in H_2 and N_2 from transport coefficients, Phys. Rev., 127:1621.

Gee, N., and Freeman, G. R., 1986, Geminate recombination of electrons in liquid methane, J. Chem. Phys., 85:1206.

Green, A. E. S., and Sawada, T., 1972, Ionization cross sections and secondary electron distributions, J. Atmos. Terrest. Phys., 34:1719.

Hake, R. D., and Phelps, A. V., 1967, Momentum transfer and inelastic collision cross sections for electrons in O_2, CO, and CO_2," Phys. Rev., 158:70.

Kline, L. E., 1982, Performance predictions for electron-beam controlled on/off switches, IEEE Trans. on Plasma Sci., 10:224.

Langevin, P., 1903, Ann. de Chim. et de Phys., 28:433.

Lin, S. L., and Bardsley, J. N., 1978, The null-event method in computer simulation, Comput. Phys. Commun., 15:161.

Littlewood, I. M., Cornell, M. C., Clark, B. K., and Nygaard, K. J., 1983, Two- and three-body electron-ion recombination in carbon dioxide, J. Phys. D, 16:2113.

Loeb, L. B., 1955, "Basic Processes in Gaseous Electronics," University of California Press, Berkeley.

Massey, H. S. W., and Gilbody, H. B., 1974, "Electronic and Ionic Impact Phenomena IV," Clarendon Press, Oxford.

Morgan, W. L., and Bardsley, J. N., 1983, Monte Carlo simulation of electron-ion recombination at high pressure, Chem. Phys. Lett., 96:93.

Morgan, W. L., 1984a, Electron-ion recombination in water vapor, J. Chem. Phys., 80:4564.

Morgan, W. L., 1984b, Molecular dynamics simulation of electron-ion recombination in a nonequilibrium, weakly ionized plasma, Phys. Rev A, 30:979.
Morgan, W. L., 1986, Molecular dynamics simulation of geminate recombination by electrons in liquid methane, J. Chem. Phys., 84:2298.
Morgan, W. L., Bardsley, J. N., Lin, J., and Whitten, B. L., 1982, Theory of ion-ion recombination, Phys. Rev. A, 26:1696.
Nakamura, Y., Shinsaka, K., and Hatano, Y., 1983, Electron mobilities and electron-ion recombination rate constants in solid, liquid, and gaseous methane, J. Chem. Phys., 78:5820.
Onsager, L., 1938, Initial recombination of ions," Phys. Rev., 54:554.
Percival, I. C., 1982, Collisions of charged particles with highly excited atoms, in: "Atomic and Molecular Collision Theory," F. A. Gianturco, ed., Plenum Press, New York.
Pitchford, L. C., ONeil, S. V., and Rumble, J. R., 1981, Extended Boltzmann analysis of electron swarm experiments, Phys. Rev. A, 23:294.
Scofield, P., 1973, Computer simulation studies of the liquid state, Comput. Phys. Commun., 5:17.
Sennhauser, E. S., and Armstrong, D. A., 1978, Ion neutralization rates in gaseous ammonia, Radiat. Phys. Chem., 11:17.
Sennhauser, E. S., Armstrong, D. A., and Warman, J. M., 1980, The temperature dependence of three-body electron ion recombination in gaseous H_2O, NH_3, and CO_2, Radiat. Phys. Chem., 15:479.
Thomson, J. J., 1924, Phil. Mag., 47:337.
Warman, J. M., 1981, Estimates of electron thermalization times for dielectric liquids from drift velocity data, Radiat. Phys. Chem.,17:21
Warman, J. M., Sennhauser, E. S., and Armstrong, D. A., 1979, Three body electron-ion recombination in molecular gases, J. Chem. Phys., 70:995.
Whitten, B. L., Morgan, W. L., and Bardsley, J. N., 1983, Mutual neutralization in rare gas halides, J. Chem. Phys., 78:1339.
Ziman, J. M., 1979, "Models of Disorder," Cambridge University Press, Cambridge.

MACROSCOPIC AND MICROSCOPIC PERSPECTIVES OF TERMOLECULAR ASSOCIATION OF ATOMIC REACTANTS IN A GAS

M. Raymond Flannery

School of Physics
Georgia Institute of Technology
Atlanta, Georgia 30332

1. MACROSCOPIC TREATMENT

The sequence normally invoked to explain the termolecular association process

$$A + B + M \rightarrow AB + M \qquad (1.1)$$

where the associating pair (A-B) may be a positive ion-negative ion or a positive ion-neutral pair, is the *energy-transfer sequence* characterized in macroscopic terms as

$$A + B \underset{k_{-1}}{\overset{k_1}{\rightleftarrows}} AB^* \; (R \leq R_T) \qquad (1.2a)$$

$$AB^*(R \leq R_T) + M \overset{k_s}{\rightarrow} AB + M \qquad (1.2b)$$

wherein intermediate complexes AB^* formed with internal separations $R \leq R_T$ at rate k_1 ($cm^3\ s^{-1}$) can decompose naturally at frequency k_{-1} (s^{-1}) or can be stabilized at rate k_s ($cm^3\ s^{-1}$) via energy-changing collisions with the atomic or molecular species M of a thermal gas. The inverse of this sequence is known as the *Lindemann mechanism* (cf. Forst 1973) for collisional dissociation.

Process (1.1) may also proceed in parallel by the *chaperone mechanism*

$$A + M + (M) \underset{k_d}{\overset{k_c}{\rightleftarrows}} AM^* + (M) \qquad (1.3a)$$

$$AM^* + B \overset{k_r}{\rightarrow} AB + M \qquad (1.3b)$$

wherein intermediate complexes AM^* which may include bound levels are formed at rate k_c ($cm^3\ s^{-1}$) by two- or three-body collisions and then undergo rearrangement with B at rate k_r ($cm^3\ s^{-1}$). The function of the chaperone M is not to keep the reactants A and B apart but to promote their stable union. Since the dependence on gas density of the resulting rate laws for formation of AB are similar, distinction between the two parallel processes is not possible from measurements based on rate laws. The energy transfer sequence (1.2) is dominant for *termolecular ion-ion recombination* owing to the long-range Coulombic attraction and is normally invoked for *ion-molecule association*. The chaperone mechanism (1.3) can however become important for association, particularly at lower gas temperatures and higher gas densities, and in some instances can be dominant (Blades and Kebarle, 1983 for COH^+-CO and N_2H^+-N_2 association in H_2).

The effective two-body rate $\alpha(cm^3\ s^{-1})$ or the associated termolecular rate k_T ($cm^6\ s^{-1}$) for (1.1) via the sequence (1.2) is

$$\alpha n_A n_B = k_s n^*_{AB}\ N = k_T\ N \tag{1.4}$$

where the gas density is N, and where the density n^*_{AB} of complex AB^* pairs satisfies

$$dn^*_{AB}/dt = k_1\ n_A n_B - (k_{-1} + k_s N)n^*_{AB} \tag{1.5}$$

where n_A and n_B are the densities of the reactants A and B. In steady-state

$$n^*_{AB}/n_A n_B = k_1/(k_{-1}+k_s N) = k_1[K/(k_1+k_s KN)] \tag{1.6}$$

where the *reaction volume* under full equilibrium conditions is

$$K = \tilde{n}^*_{AB}/\tilde{n}_A\tilde{n}_B = k_1/k_{-1} \tag{1.7}$$

which is not an equilibrium constant in the usual sense since the complex AB^* does <u>not</u> include bound states. The overall rate

$$\alpha(N) = \frac{k_1(K\ k_s N)}{k_1+(Kk_s N)} \rightarrow \begin{cases} K\ k_s N = \alpha_o & ;\ N \rightarrow 0 \\ k_1 & ;\ N \rightarrow \infty \end{cases} \tag{1.8}$$

initially increases linearly with gas density N as $\alpha_o = (Kk_s N)$, characteristic of three-body kinetics, and eventually tends to a saturation value k_1 characteristic of two-body kinetics alone. Introduce the N-dependent probabilities,

$$P^S(N) = (k_s KN)/(k_1+k_s KN) \tag{1.9}$$

for collisional stabilization of the complex and

$$P^D(N) = k_1/(k_1+k_s KN) \tag{1.10}$$

for natural decomposition of the complex so that alternative expressions for the effective two-body rate are

$$\alpha = P^S(N)k_1 \tag{1.11}$$

$$= P^D(N)\ k_s KN \equiv P^D(N)\alpha_o \tag{1.12}$$

1.1. The Reaction Volume

By virtue of (1.8), K represents the reaction volume at low N and is in general given by (Forst 1973)

$$K(T) = \frac{h^3}{(2\pi mkT)^{3/2}} \frac{q(AB^*)}{q(A)q(B)} \frac{\omega_{AB}}{\omega_A\omega_B} \tag{1.13}$$

where q and ω denote the internal partition functions and electronic statistical weights associated with each reactant A and B and with the activated complex AB^*. While q(A) and q(B) are known, $q(AB^*)$ is such that it must include only those rotational-vibrational states of AB^* accessible at energies E above the dissociation threshold E_T in A-B collisions. It may also include conservation of total angular momentum which can be redistributed between the orbital angular momentum $\underset{\sim}{L}$ for relative (A-B) motion and the combined internal momentum $\underset{\sim}{J}_R$ of the individual reactants.

For structureless reactants A and B with relative separation $\underset{\sim}{R}$, relative energy $E \geq 0$ as measured above the threshold energy E_T for dissociation from the ground vibrational state of AB, relative momentum $\underset{\sim}{P}$ and angular momentum $L = \hbar J$, then

$$K(T) = h^3(2\pi mkT)^{-3/2} \int \exp(-E/kT)d\underset{\sim}{R}\ d\underset{\sim}{p}/h^3 \tag{1.14a}$$

$$= h^3(2\pi mkT)^{-3/2} \int_0^\infty dE \int_0^{J_{max}} \rho(E,J)\ \exp(-E/kT)dJ^2 \tag{1.14b}$$

The density of ro-vibrational states per unit interval dE dJ^2 produced from the relative (A-B) motion is

$$\rho(E,J) = \tau(E,J)/h = 1/h\nu \tag{1.15}$$

in terms of the lifetime

$$\tau(E,\rho) = \oint_{R_1}^{R_T} \frac{dR}{v_R} = \left(\frac{m}{2}\right)^{1/2} \oint_{R_1}^{R_T} \left[E\left(1 - \frac{\rho^2}{R^2}\right) - V(R)\right]^{-1/2} dR \tag{1.16}$$

of the complex towards natural decomposition with radial speed v_R via an orbit with angular momentum L and impact parameter ρ which are related by

$$L = \hbar J = (2mE)^{1/2}\,\rho \qquad (1.17)$$

The maximum angular momentum L_{max} of those (A^+-B) orbits which can overcome the centrifugal barriers for $L \leq L_{max}$ at fixed energy E is given by $(8m^2\alpha_B e^2 E)^{1/4}$ for pure polarization attraction $V(R) = -\alpha_B e^2/2R^4$ where α_B is the polarizability of B. The associated reaction radius R_T is taken as the barrier location $(\alpha_B e^2/2E)^{1/4}$, the Langevin orbiting radius.

So as to obviate the necessity for numerical evaluation of the time (1.16) during which R is in the range $R_1 \leq R \leq R_T$ for realistic interactions V(R), a simplifying assumption is now invoked. Reactions which occur via intermediate complexes generally involve randomization of the internal energy in the complex. If this energy is statistically distributed within the various internal modes the reaction can be treated by statistical methods such that the (formidable) problems associated with application of molecular dynamics and classical trajectories to microscopic lifetimes of intermediate complexes are thereby avoided. The complex is therefore assumed to be in its most stable configuration which results from rapid energy transfer out of the external collision coordinates $\underset{\sim}{R}$ and $\underset{\sim}{p}$ into other internal coordinates so that, following large perturbations in (A^+-B) close encounters, a quasi-equilibrium among excited states is obtained. Upon this key randomization assumption (which is viable provided the lifetime against stabilization collisions is long in comparison to the (A-B) collision time) the density (1.15) may therefore be replaced by the density

$$\rho_H(E,J) = 1/h\nu_H(E,J) \qquad (1.18)$$

of levels (one per energy-level spacing $h\nu_H$) of a harmonic oscillator of frequency ν_H. The number of levels within rotational spacing dJ of the free rotor complex remains as $dJ^2 = 2J\,dJ$. The difficult molecular-dynamics aspects of the complex have therefore been avoided and are replaced by the known structural properties of the complex. Eq. (1.14b) with (1.18) can be formally generalized to cover polyatomic reactants.

1.2 Temperature Dependence: Polyatomic Reactants

Temperature dependence of association reactions elucidate reaction pathways. Since the unimolecular dissociation rate k_{-1} increases with T, a T^{-n}-variation is expected. Current theories have mainly focused on the temperature dependence of the low density three body rate $k_T = Kk_s$. Accord with measurements is becoming somewhat acceptable (Bates 1985, Viggiano 1985, van Koppen et al. 1984). The rate k_s for stabilization of ion-molecule collision complexes may be taken as some fraction of the constant Langevin rate for spiraling collisions,

$$k_L = 2\pi e(\alpha_M/\mu)^{1/2} \qquad (1.19)$$

where α_M is the polarizability of M and where μ is the reduced mass of the (AB^*-M) system. Since the partition functions q(A) and q(B) of the reactants are known, (e.g., q(A)q(B) for ground vibrational levels

$\sim T^{-(r/2)}$ where r is the sum (r_A+r_B) of the number of degrees of rotational freedom of the isolated reactants), the temperature dependence of k_T is therefore mainly controlled by the less well determined partition function $q(AB^*)$ in (1.13) of the activated complex. The different temperature dependences exhibited by similar systems, e.g., k_T for $(O_2^+-O_2)-O_2$ association decreases as $\sim T^{-2.9} - T^{-3.2}$ which is more rapid than $k_T \sim T^{-1.6} - T^{-1.85}$ for $(N_2^+-N_2) - N_2$ association, is simply a manifestation (Bates 1985) of the temperature characteristics of $q(AB^*)$.

Bates (1978) has outlined a scheme for the internal partition function $q(AB^*)$ which is the generalization of (1.14a) and of (1.15) to atomic-diatomic and to diatomic-diatomic ion-neutral reactants which are free to rotate and vibrate in the potential well. The states accessible in Langevin collisions are determined by conservation of energy and angular momentum. As with (1.14b), this *barrier model* of Bates in general involves time-consuming numerical calculation of the lifetime τ for A-B collisions. Upon invoking the randomization assumption, however, this lifetime may be replaced, as in (1.18), by known vibrational periods of the complex.

Apart from the long-range polarization (A^+-B) attraction and the resulting looseness of the activated complex $(AB^+)^*$, termolecular ion-molecule association as described by (1.2) is in principle the inverse of neutral unimolecular decomposition which is primarily concerned with calculation of the decomposition rate k_{-1}. The Rice-Ramsperger-Kassel-Marcus (RRKM) theory (cf. Forst 1973) for calculation of the internal partition function of the complex has been adapted for ion-neutral fragmentation. In his *thermal RRKM method* for ion-molecule association Herbst (1979) includes all ro-vibrational states above the dissociation limit at E_T. Following the *barrier model* of Bates (1979a,b) it was subsequently modified (Herbst 1980) to include only those levels above the centrifugal barrier. The *quasi-equilibrium theory* QET (Klots 1971) and *phase space theory* PST (Bass et al. 1979), in contrast to RRKM, rigorously conserve angular momentum of all states of the system including relative orbital and internal rotational angular momentum of the fragments and accounts for the centrifugal barrier. Marcus (1975) has modified "tight" transition state theory to cover the cases when the total angular momentum of the complex is transferred totally to orbital or totally to internal angular momenta of the products. The effect of internal angular momenta of ion-neutral reactants is however minor and can be ignored (Bates 1985) in association.

The generalization of (1.14b) to polyatomic reactants together with the appropriate generalization of the randomization assumption (1.18) yields the result of phase space theory (PST). By using PST and the practical expressions of Troe (1977), Bates (1986) provided a result which is identical with (1.14b) but with ρ therein replaced by

$$\rho(E,J) = \rho_V(E+E_T)\ F(E_T,J)/\sigma \qquad (1.20)$$

where σ is the (symmetry) number of indistinguishable ways of orienting the AB^*-complex. The density of vibrational states at energy E above the dissociation threshold $E_T = [E_o+(1/2)\sum_{i=1}^{s} h\nu_i] \equiv E_0+E_Z$ as measured from the minimum of the potential energy surface V(R) is,

$$\rho_V(E+E_T) = [(s-1)/(s-1.5)]^m (E+E_T)^{s-1}/[(s-1)! \prod_{i=1}^{s} h\nu_i] \qquad (1.21)$$

where ν_i are the frequencies of s harmonic oscillators in $(AB^+)^*$, and where m is the number of oscillators that disappear during dissociation. The first factor [ ] in (1.21) is an anharmonic correction, and for complexes with more than three atoms, $E_T = E_0 + aE_Z$ where a is the correction of Whitten and Rabinovitch (1963) designed to improve the semiclassical approximation involved with the zero point energy. The factor F in (1.20) is

$$F(E_T,J) = \gamma[1 - BJ^2/E_T]^{s-1} \qquad (1.22)$$

(γ = 1, and 2J for linear and symmetrical-top complexes, respectively) and acknowledges the effect that the averaged rotational energy BJ^2 of the complex has reduced the energy available to vibrational modes. For structureless reactants s is unity so that (1.21) reproduces the original result (1.14b) with (1.15).

1.3. Density Dependence

In contrast to the temperature dependence, the density dependence (1.8) or its microscopic equivalent (§ 2) has been tacitly accepted without any critical analysis. In order to isolate the key issue, express (1.14) in the equivalent form

$$K(T) = \int_0^\infty v_\infty\, G(E)dE \int_0^{\rho_{max}} \tau(E,\rho)(2\pi\rho d\rho) \qquad (1.23)$$

where the dynamical lifetime τ is given by (1.16), where $E = mv_\infty^2/2$ and where $\rho_{max} = R_T[1-V(R_T)/E_i]^{1/2}$ is the maximum impact parameter ρ of that orbit which just gains entry to the R_T-sphere. For polarization attraction $\pi\rho_{max}^2$ is $2\pi e(\alpha_B/m)^{1/2}\, v_\infty^{-1}$ such that the thermal approach rate $k_1 = \pi\rho_{max}^2 v_\infty$ is independent of T.

When the (A-B) interaction V(R) is neglected, τ is $2(R_T^2 - \rho^2)^{1/2}/v_\infty$, and (1.23) for K reduces naturally to the volume $4\pi R_T^3/3$ of the complex. For termolecular ion-ion recombination the stabilization frequency $\nu_S = k_S N$ can be separated as $\langle v\rangle(\lambda_A^{-1} + \lambda_B^{+1})$ where $\lambda_{A,B}$ are the path lengths for individual A-M and B-M binary collisions at average speed $\langle v\rangle$. The stabilization probability (1.9) for $\lambda_A = \lambda_B = \lambda$ is then

$$P_K^S(N) = \frac{(8X/3)}{1+(8X/3)} = (8X/3)\left[1 - (8X/3) + (8X/3)^2 - (8X/3)^3 \ldots\right], \qquad (1.24)$$

by the *partition-function method*, where X is R_T/λ.

The stabilization probability in (1.11) may also be identified as

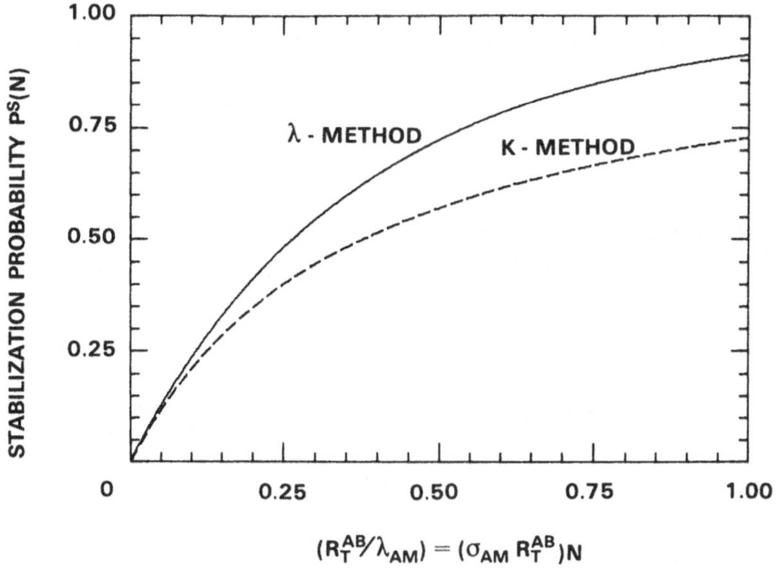

Fig. 1. $P^S(N)$ as determined by free-path (λ) and partition (K) methods.

$$P_T^S(N) = W(X_A) + W(X_B) - W(X_A)W(X_B) \tag{1.25}$$

the combined probability for A-M and B-M collisions. The Thomson probability W for individual A-M collision for rectilinear (A-B) relative motion within the R_T-sphere is explicitly determined (Loeb 1955) as

$$W(X) = 1 - \frac{1}{2X^2}[1-\exp(-2X)(1+2X)] = \sum_{n=1}^{\infty} (-1)^{n+1} \frac{2^{n+1}X^n}{(n+1)!}\frac{(n+1)}{n+2} \tag{1.26}$$

so that the total probability (1.25) for AB^* $(R\leq R_T)$-M stabilization encounters is

$$P_T^S(N) = (8X/3)\,[1 - (17X/12) + (7X^2/5) - (129X^3/120) + \ldots\,] \tag{1.27}$$

Although (1.24) and (1.27) agree in the low and high density limits, there is significant departure at intermediate densities (Figure 1). For ion-atom association the appropriate comparison is between the single collision probability $W(X = R_T/\lambda_{AB})$ for AB^*-M collisions with free path λ_{AB} and $(4X/3)/(1+4X/3)$, the corresponding P_K^S, and is similar to that depicted in Fig. 1. For association note that M does not necessarily have to be within the R_T-sphere for collisions to occur.

Dilemma 1: Which of the above methods is more accurate? The outcome is certainly important in principle and may be important in practice since (1.8) is generally used to deduce third order rates α_o from measurements at various N (cf. Mahan and Person 1964 for recombination and Bates 1986 for association). For ion-ion recombination the Thomson radius (see §4) R_T is $(2/3)e^2/kT \sim 370Å$ at 300 K, and for ion-atom association the Langevin radius R_L of the AB^+ complex is $(\alpha_B e^2/3kT)^{1/4} \sim (5\text{-}15)Å$. Hence the range of X in Fig. 1 which corresponds to gas pressures up to (1/2) atm. (where $\lambda \sim 180Å$ at STP) is $0 \rightarrow 1$ for recombination and $0 \rightarrow \sim 0.05$ for association. For most pressures of practical interest, therefore, variation of the association rate is given by either method (cf. Fig. 1). For recombination, there is a large difference. For higher pressures, however, diffusional drift (§4) ignored by both methods becomes important only for recombination.

Dilemma 2: Although the form of (1.8),

$$\alpha = (k_1\alpha_o)/(k_1+\alpha_o) \quad , \qquad \alpha_o = Kk_TN \tag{1.28}$$

is aesthetically pleasing in that it illustrates quite naturally that the overall rate is given by the rate limiting step, k_1 being the thermal rate as such does not acknowledge diffusional drift (§4) which arises from a non-equilibrium distribution in R outside the reaction sphere. As the gas density N is raised can we simply replace the thermal rate k_1 of approach by the rate α_{TR} of transport by diffusional drift to R_T?

In order to address these queries a microscopic treatment of the energy-transfer sequence (1.2) for structureless reactants in a chemically inert gas must first be developed (§ 2). Appropriate generalization to structured reactants, if required, appears complicated but feasible.

2. MICROSCOPIC DEVELOPMENT FOR ATOMIC REACTANTS

The set of Master Equations derived from the Boltzmann equation for the two-particle distribution function (Flannery 1982) for the (cm^{-3}) concentrations $n_i^{\pm}(\underset{\sim}{R},t) \equiv n_i^{\pm}(\underset{\sim}{R},E_i,L_i^2;t)$ per unit interval $d\underset{\sim}{R}\, dE_i\, dL_i^2$ of A-B pairs with internal energy E_i, and internal angular momentum (squared) L_i^2 that are internally expanding (+) and contracting (-) with radial speed $|v_R|$ at internal separation R is (Flannery 1986, 1987)

$$dn_i^{\pm}(\underset{\sim}{R},t)/dt = \partial n_i^{\pm}(\underset{\sim}{R},t)/\partial t \pm \frac{1}{R^2}\frac{\partial}{\partial R}\left[R^2\, n_i^{\pm}(\underset{\sim}{R},t)\,|v_R|\right]_{E_i,L_i^2}$$

$$= -\int_{V_i(R)}^{\infty} dE_f \int_0^{L_{mf}^2} S_{if}^{\pm}(\underset{\sim}{R},t)\, dL_f^2 \tag{2.1}$$

where the net collisional rate per unit intervals $dE_i dL_i^2$ and $dE_f dL_f^2$

about the initial and final states i and f respectively is

$$S_{if}^{\pm}(\underset{\sim}{R},t) = n_i^{\pm}(\underset{\sim}{R},t)\ \nu_{if}(R) - n_f^{\pm}(\underset{\sim}{R},t)\ \nu_{fi}(R) = -S_{fi}^{\pm}(\underset{\sim}{R},t) \tag{2.2}$$

in terms of the frequency $\nu_{if}(R)$ per unit interval $dE_f dL_f^2$ for transitions $i\ (E_i,L_i^2) \to f\ (E_f,L_f^2)$ induced by collisions between M and the pair (A-B). The relative interaction energy between A and B of reduced mass m is V(R) and the effective radial interaction is

$$V_i(R) = V(R) + L_i^2/2mR^2 \tag{2.3a}$$

with a barrier height U_i at location R_i.

For fixed E_i and R, the maximum value of L_i^2 is

$$L_{mi}^2(E_i,R) = 2m\ [E_i - V(R)]R^2 \tag{2.3b}$$

and, for fixed E_i alone, the maximum value of L_{mi}^2 with respect to R is $L_{oi}^2(E_i)$.

The basic expression for the rate ($cm^3\ s^{-1}$) of termolecular association processes (1.1) at time t is (Flannery 1985)

$$R^A(t) = \int_{-D}^{\infty} dE_i \int_o^{L_{oi}^2} dL_i^2 \int_{R_i^-}^{R_i^+} P_i^A(E_i,L_i^2;R) \left[\frac{dn_i(\underset{\sim}{R},t)}{dt}\right] d\underset{\sim}{R} \tag{2.4}$$

where the summed distribution $n_i = n_i^+ + n_i^-$ of the expanding and contracting R-pairs in (2.1) satisfies (Flannery 1986)

$$dn_i(\underset{\sim}{R},t)/dt = \partial n_i(\underset{\sim}{R},t)/\partial t + \frac{1}{R^2}\frac{\partial}{\partial R}\left[R^2(n_i^+ - n_i^-)|v_R|\right]_{E_i,L_i^2}$$

$$= -\int_{V_i}^{\infty} dE_f \int_o^{L_{mf}^2} S_{if}(\underset{\sim}{R},t)dL_f^2 \tag{2.5}$$

in which S_{if} is the sum $(S_{if}^+ + S_{if}^-)$ of collision rates defined in (2.2).

In (2.4), P_i^A is the probability that (E_i,L_i^2) -pairs are in the product (association) channel, R_i^- and R_i^+ are the pericenter and apocenter of the relative (E_i,L_i^2) motion, and -D is the lowest vibrational energy level of AB with respect to the dissociation limit taken as zero energy.

(A) Assume that pairs in the block $\mathcal{E}$ of highly excited bound AB levels, sandwiched between the continuum block $\mathcal{C}$ ($E_i > U_i$, all R; $0 \le E_i \le U_i$, $R \ge R_i$) of fully dissociated states (where P_i^A is zero) and the sink block $\mathcal{S}$ of fully associated stabilized levels between -D and some level -S (throughout which P_i^A is unity), are in quasi-steady-state (QSS). Hence

$$n_i(t) = \int_{R_i^-}^{R_i^+} n_i(\underset{\sim}{R},t)d\underset{\sim}{R} \tag{2.6}$$

in block $\mathcal{E}$ satisfies

$$dn_i/dt = \partial n_i/\partial t = -\int_{R_i^-}^{R_i^+} d\underset{\sim}{R} \int_{V_i}^{\infty} dE_f \int_0^{L_{mf}^2} S_{if}(\underset{\sim}{R},t)dL_f^2 \approx 0 \tag{2.7}$$

since $n_i^+ = n_i^-$ at both turning points $R_i^{(\pm)}(E_i,L_i^2)$. Under the conservation requirement

$$\frac{\partial}{\partial t}\int_{-D}^{\infty} dE_i \int_0^{L_{oi}^2} n_i(t)dL_i^2 = \frac{\partial}{\partial t}\int_0^{\infty} dL_i^2 \int_{V_i}^{\infty} n_i(t)dE_i = 0 \tag{2.8}$$

for a complete steady-state system, (2.4) therefore reduces under QSS of block $\mathcal{E}$, to the $\mathcal{E} \to \mathcal{S}$ current or $\mathcal{C} \to \mathcal{E}$ current

$$R^A(t) = \int_{-D}^{-S} dE_i \int_0^{L_{oi}^2} dL_i^2 \int_{R_i^-}^{R_i^+} \left[\frac{dn_i}{dt}\right] d\underset{\sim}{R} = -\int_0^{\infty} dL_i^2 \int_{U_i}^{\infty} dE_i \int_{R_i^-}^{R_i} \left[\frac{dn_i}{dt}\right] d\underset{\sim}{R} \tag{2.9a}$$

where R_i is the location of the centrifugal barrier U_i of V_i. With the aid of (2.5) the $\mathcal{C} \to \mathcal{E}$ current is

$$R^A(t) = \int_0^{\infty} dL_i^2 \int_{U_i}^{\infty} dE_i \; 4\pi R_i^2 \; |v_R(R_i)| \; [n_i^-(R_i)-n_i^+(R_i)] \tag{2.9b}$$

the net inward transport of pairs across R_i or

$$R^A(t) = \int_0^{\infty} dL_i^2 \int_{U_i}^{\infty} dE_i \int_{R_i^-}^{R_i} d\underset{\sim}{R} \int_{V_i}^{U_i} dE_f \int_0^{L_{mf}^2} S_{if}(\underset{\sim}{R},t)dL_f^2 \tag{2.9c}$$

the net collisional current flowing downwards across the necks between R_i^- and R_i provided by the centrifugal barrier of height U_i.

2.1 Rate of Association of Activated Complex

The density (per unit interval $d\Gamma_i = dE_i dL_i^2$)

$$n_i^*(t) = \int_{R_i^-}^{R_T} n_i(\underset{\sim}{R},t)d\underset{\sim}{R} \tag{2.10}$$

of (E_i,L_i^2)-pairs which are assumed to form an activated complex with internal separation R in the range $R_i^- \leq R \leq R_T$ (E_i,L_i^2) satisfies

$$\partial n_i^*(t)/\partial t = 4\pi R_T^2\, v_R(R_T)n_i^-(R_T)[1-P_i^d(R_T)] - \int_{R_i^-}^{R_T} d\underset{\sim}{R} \int S_{if}(\underset{\sim}{R},t)d\Gamma_f \tag{2.11a}$$

$$= 4\pi R_T^2 v_R(R_T)n_i^-(R_T)P_i^s(R_T) - \int_{R_i^-}^{R_T} d\underset{\sim}{R} \int S_{if}(\underset{\sim}{R},t)d\Gamma_f \tag{2.11b}$$

where the implied two-dimensional grid for Γ_f-integration spans the range $0 \leq L_f^2 \leq L_{of}^2$ and $V_i \leq E_f \leq U_i$. The microscopic probabilities

$$P_i^d(R_T) = n_i^+(R_T)/n_i^-(R_T) \tag{2.12}$$

for decomposition of the activated (E_i,L_i^2,R_T)-complex, and

$$P_i^s(R_T) = [n_i^-(R_T)-n_i^+(R_T)]/n_i^-(R_T) = 1 - P_i^d(R_T) \tag{2.13}$$

for collisional stabilization of the complex, are introduced in (2.11).

Under the assumption that only those pairs within the complex can participate in association, the basic QSS-rate (2.9b) for association in terms of the stabilization probability (2.13) is now

$$R^A(t) = \alpha n_A n_B = \int_0^\infty dL_i^2 \int_{U_i}^\infty dE_i\; [4\pi R_T^2 v_R(R_T)n_i^-(R_T)]P_i^s(R_T) \tag{2.14}$$

in a form suitable for a general choice of $R_T(E_i,L_i^2)$. The radius R_T may be identified with the location R_i of the centrifugal barrier. When R_T is constant, or is only a function of E_i, then the equivalent form

$$\alpha n_A n_B = \int_0^\infty dE_i \int_0^{L_{max}^2} dL_i^2\; [4\pi R_T^2 v_R(R_T)n_i^-(R_T)]P_i^s(R_T) \tag{2.15}$$

is also appropriate, where

$$L_{max}^2 = 2m\, E_i[1 - V(R_T)/E_i]R_T^2 = (2mE_i)\rho_{max}^2 \tag{2.16}$$

is associated, as in §1, with that limiting orbit with impact parameter $\rho = \rho_{max}$ which just touches the R_T-sphere.

2.2 Treatment with Thermal Rate of Complex Formation

When there is a thermal distribution of AB-pairs with internal separation $R \geq R_T$

$$n_i^-(\underset{\sim}{R})/n_A n_B = \tilde{n}_i^-(\underset{\sim}{R})/\tilde{n}_A\tilde{n}_B \tag{2.17}$$

where the tildas (~) signify thermodynamic equilibrium, then (2.14) yields,

$$\alpha = \int_0^\infty dL_i^2 \int_{U_i}^\infty dE_i\; k_i(R_T)P_i^s(R_T) \tag{2.18}$$

in which the one-way thermal equilibrium rate k_i for formation of R_T-pairs (per unit interval $dE_i dL_i^2$) may be written in the following variety of suggestive forms,

$$k_i = 4\pi R_T^2 v_R(R_T)\tilde{n}_i^-(\underset{\sim}{R}_T)/\tilde{n}_A\tilde{n}_B \tag{2.19a}$$

$$= 4\pi^2 \exp(-E_i/kT)/(2\pi mkT)^{3/2} \tag{2.19b}$$

$$= \pi R_T^2[1 - V(R_T)/E_i]v_\infty\; G(E_i)/L_{max}^2 \tag{2.19c}$$

$$= [\tilde{n}_i^*(E_i,L_i^2)/\tilde{n}_A\tilde{n}_B]/\tau_i^d \tag{2.19d}$$

$$= v_\infty\; G(E_i)\; 2\pi\; \rho(d\rho/dL_i^2) \tag{2.19e}$$

each being appropriate to application of (2.18) or (2.15) with (2.17). The Maxwellian distribution of relative energies $E_i = m\, v_\infty^2/2$ is

$$G(E_i)dE_i = 2(E_i/\pi kT)^{1/2} \exp(-E_i/kT)\; d(E_i/kT) \tag{2.20}$$

For the polarization (A^+-B) potential, use of (2.15) involves $R_T(E_i) = (\alpha_B e^2/2E_i)^{1/4}$ and $L_{max} = (8m^2\alpha_B e^2 E_i)^{1/4}$, as in §1, and use of (2.18) involves $R_T(L_i) = (2\alpha_B me^2)^{1/2}/L_i$ and $U_i(R_T) = \alpha_B e^2/2R_T^4$, the barrier height.

The equilibrium microscopic concentration per unit interval dE_i of <u>all</u> pairs in the $R \leq R_T$ complex in (2.19d) is

$$\tilde{n}_i^*(E_i,L_i^2)dL_i^2 = dL_i^2 \int_{R_i^-}^{R_T} \tilde{n}_i(\underset{\sim}{R},E_i,L_i^2)d\underset{\sim}{R} = \tilde{n}_A\tilde{n}_B[v_\infty G(E_i)][2\pi\rho d\rho]\tau_i^d \tag{2.21}$$

where the natural lifetime of the complex is

$$\tau_i^d(E_i,L_i^2) = \int_{R_i^-}^{R_T} dR/v_R = [\nu_i^d(E_i,L_i^2)]^{-1} \tag{2.22}$$

Note that the reaction volume $K_i = \tilde{n}_i^*/\tilde{n}_A\tilde{n}_B$ per unit $dE_i dL_i^2$, the microscopic equivalent of (1.14), may be written in various forms as

$$K_i(E_i,L_i)dE_i dL_i^2 = h^3(2\pi mkT)^{-3/2}\ \rho_i(E_i,J_i)\ \exp(-E_i/kT)dE_i dJ_i^2 \tag{2.23a}$$

$$= (\pi/2mE_i)[v_\infty G(E_i)dE_i]dL_i^2 \tag{2.23b}$$

$$= [v_\infty G(E_i)dE_i][2\pi\rho d\rho]\tau_i^d(E_i,\rho) = k_i\tau_i^d\ dE_i dL_i^2 \tag{2.23c}$$

where the density of states ρ_i is $(h\nu_i^d)^{-1}$, which is the microscopic generalization of (1.15).

Provided that the stabilization probability P_i^S can be determined, the rate α follows directly from (2.18). Common to the following two approximations for P_i^S is the *strong-collision assumption* that collisional absorption within the complex only occurs. Thus, for a steady state distribution of pairs in the complex, (2.1) is simplified to yield,

$$\pm \frac{1}{R^2}\frac{\partial}{\partial R}[R^2 n_i^{\pm}(R)|v_R|]_{E_i,L_i^2} = -\ n_i^{\pm}(\underset{\sim}{R})\nu_i(R) \tag{2.24}$$

where the frequency for collisional formation of bound pairs with $R \leq R_T$ below the centrifugal barrier is

$$\nu_i(R) = \int_{V_i}^{U_i} dE_f \int_0^{L_{mf}^2} \nu_{if}(R)dL_f^2 \tag{2.25}$$

2.3 Partition-Function Method for Stabilization Probability

This first method for calculation of P_i^S is based on (2.11) which is the microscopic analogue of (1.5). On defining the one-way equilibrium microscopic rate of formation or thermal approach rate as in §2.1 by

$$k_i\ \tilde{n}_A\tilde{n}_B = 4\pi R_T^2 v_R(R_T)\tilde{n}_i^-(\underset{\sim}{R}_T)\ , \tag{2.26}$$

and the non-equilibrium rates for natural decomposition,

$$k_i^* n_i^* = 4\pi R_T^2 v_R(R_T) n_i^+(\underset{\sim}{R}_T) \quad , \tag{2.27}$$

and for collisional stabilization

$$\nu_i^s n_i^* \equiv (k_i^s N) n_i^* = \int_{R_i^-}^{R_T} n_i(R)\nu_i(R) d\underset{\sim}{R} \tag{2.28}$$

of the complexes, then (2.11) can be expressed as

$$\partial n_i^* / \partial t = k_i n_A n_B - (k_i^* + k_i^s N) n_i^* \tag{2.29}$$

For a steady state distribution of complexes, the rate (2.18), with the aid of (2.25) - (2.29), is

$$R^A(t) = \int_0^\infty dL_i^2 \int_{U_i}^\infty dE_i \; [n_i^* \; k_i^s \; N] \tag{2.30}$$

$$= \int_0^\infty dL_i^2 \int_{U_i}^\infty dE_i \left[\frac{k_i(k_i^s K_i N)}{(k_i + k_i^s K_i N)}\right] \tag{2.31}$$

which is the microscopic generalization of (1.8). The reaction volume

$$K_i = \tilde{n}_i^*(E_i, L_i^2)/\tilde{n}_A \tilde{n}_B = k_i / k_i^* \tag{2.32}$$

per unit $dE_i dL_i^2$ is given by any of the forms in (2.23), whichever proves most convenient. The integration order in (2.31) may be interchanged as in (2.15). The form of (2.31) is well known (Bass et al. 1979, Bates 1986).

With the aid of (2.23) the ratio

$$K_i / k_i = \tau_i^d = 1/k_i^* \tag{2.33}$$

is simply the lifetime of the (E_i, L_i^2)-complex towards natural decomposition. Hence,

$$\alpha = \int_0^\infty dL_i^2 \int_{U_i}^\infty dE_i \; [k_i \; P_i^s(N)] = \int_0^\infty dL_i^2 \int_{U_i}^\infty dE_i \; [k_i^s \; K_i N \; P_i^d(N)] \tag{2.34}$$

where k_i is given by (2.19), and where the probabilities for collisional stabilization and natural decomposition can be identified with the aid of (2.28), as

$$P_i^s(N) = k_i^s K_i N/(k_i + k_i^s K_i N)$$

$$= \nu_i^s \tau_i^d/(1 + \nu_i^s \tau_i^d) \rightarrow \begin{cases} \nu_i^s \tau_i^d & , N \rightarrow 0 \\ 1 - 1/\nu_i^s \tau_i^d, & N \rightarrow \infty \end{cases} \quad (2.35)$$

and as

$$P_i^d(N) = k_i/(k_i + k_i^s K_i N)$$

$$= 1/(1 + \nu_i^s \tau_i^d) \rightarrow \begin{cases} 1 & , N \rightarrow 0 \\ 1/\nu_i^s \tau_i^d & , N \rightarrow \infty \end{cases} \quad (2.36)$$

respectively. When R_T is constant or depends only on E_i, the order of integrations in (2.34) can be interchanged and performed over the range $0 \leq E_i \leq \infty$ and $0 \leq L_i^2 \leq L_{max}^2$. The above expressions are the appropriate microscopic generalizations of (1.8)-(1.10). At low densities N therefore

$$\alpha = \int_0^\infty dL_i^2 \int_{U_i}^\infty dE_i \; k_i(E_i, L_i^2) \; [\nu_i^s \tau_i^d] = \int_0^\infty dL_i^2 \int_{U_i}^\infty K_i(E_i, L_i^2) \nu_i^s \; dE_i \quad (2.37)$$

which is $(\nu_i^s \tau_i^d)$ averaged over k_i, the one-way equilibrium rate (2.19) for complex formation, or equivalently is ν_i^s, the stabilization frequency in (2.28) averaged over the reaction volume K_i. Note that α may be expressed as $k_1 \langle \tau^d \rangle (k_s N)$ only when the so-called average lifetime $\langle \tau^d \rangle$ is defined as above (see also Bates 1979a). The low-density rate (2.37) may also be expressed as,

$$\alpha \; \tilde{n}_A \tilde{n}_B = \int_0^\infty dL_i^2 \int_{U_i}^\infty dE_i \; \tilde{n}_i^*(E_i, L_i^2) \; \nu_i^s \quad (2.38)$$

which is the equilibrium version of (2.30). When the stabilization frequency ν_i^s does not depend on (E_i, L_i^2), and is taken as a constant β times the Langevin rate (1.19) then the low density rate (2.38) reduces to

$$\alpha \; \tilde{n}_A \tilde{n}_B = \beta (k_L N) \tilde{n}^* \quad (2.39)$$

where $\tilde{n}^*$ is the concentration of pairs above all the centrifugal barriers, in accord with (1.4).

2.4. Free-path Method for Stabilization Probability

This method is based on the direct solution of (2.24) subject to the boundary conditions

$$n_i^-(R_T) = \tilde{n}_i^-(R_T)$$

$$n_i^+(R_i^-) = n_i^-(R_i^-) \quad (2.40)$$

where $R_i(E_i,L_i^2)$ is, as before, the perigee of the (E_i,L_i^2)-orbit. Hence (Flannery 1985)

$$n_i^-(R_i^- \le R \le R_T, E_i, L_i^2) = \tilde{n}_i^-(R) \exp[-\int_R^{R_T} a_i(R)dR]$$

$$n_i^+(R_i^- \le R \le R_T, E_i, L_i^2) = \tilde{n}_i^-(R) \exp[-\oint_{R_i}^{R_T} a_i(R)dR]\exp[\int_R^{R_T} a_i(R)dR] \qquad (2.41)$$

are the (R,E_i,L_i^2)-distributions of individual AB pairs within the complex.

The number of collisions or stabilizations that occur in a time $dt = dR/v_R$ or within element ds of the (E_i,L_i^2) trajectory for (A-B) relative motion at relative speed v_i are

$$a_i(R,E_i,L_i^2)dR = \nu_i dt = \nu_i dR/v_R = \nu_i ds/v_i = ds/\lambda_i \qquad (2.42)$$

where λ_i is the free path length (v_i/ν_i) between collisions at frequency ν_i given by (2.25). The stabilization probability is therefore

$$P_i^S(R_T,E_i,L_i^2) = 1 - \exp(-2\int_{R_i^-}^{R_T} ds/\lambda_i) \qquad (2.43)$$

the microscopic probability of collision of (E_i,L_i^2)-pairs with separation R in the range $R_i^- \le R \le R_T(E_i,L_i^2)$. The association rate under the strong-collision assumption is then (2.34) with (2.43).

On defining, with the aid of the stabilization frequency in (2.28),

$$\nu_i^s \tau_i^d = \oint_{R_i^-}^{R_T} \nu_i(R)dR/v_R = \oint ds/\lambda_i \qquad (2.44)$$

which equals $\nu_i^s \tau_i^d$ only if the collision frequency (2.25) in (2.28) is independent of R, then (2.43) is simply

$$P_i^S = 1 - \exp(-\nu_i^s \tau_i^d) \qquad (2.45)$$

which has been obtained from the distribution (2.41) of individual pairs within the complex. This is to be compared with the customary form (2.35) obtained from the averaged distribution in (2.29).

For straight line trajectories at constant speed and with $\nu_i = \nu_i^s = \langle v\rangle (\lambda_A^{-1} + \lambda_B^{-1})$, then τ_i^d of (2.33) and τ_i^d of (2.44) are each equal to

$2(R_T^2 - \rho^2)^{1/2}/\langle v \rangle$ so that the above free-path method for A-M and B-M collisions yields

$$P_{Ti}^S = 1 - \exp - [X_A(\rho) + X_B(\rho)] \quad , X_{A,B} = 2(R_T^2 - \rho^2)^{1/2}/\lambda_{A,B} \tag{2.46}$$

which is to be compared with the corresponding result

$$P_{Ki}^S = (X_A + X_B)/[1+(X_A+X_B)] \tag{2.47}$$

given by the partition-function method. The comparison for various impact parameters ρ illustrates discrepancies similar to that displayed in Fig. 1. It is worth noting that (2.15) with (2.43) for constant R_T and $\lambda_{A,B}$ and a straight line trajectory reduces to the rate (Flannery 1986)

$$\alpha = \pi R_T^2 \left[\int_0^\infty [1 - V(R_T)/E_i] \, v_\infty \, G(E_i) dE_i \right] P_T^S(N) \tag{2.48}$$

where the macroscopic probability P_T^S is given by (1.25). This rate is identical with (1.11) The first dilemma posed in §1 is therefore resolved. The free path method based on knowledge of individual R_T pairs within the complex represents a more basic procedure than does the partition method which is based only on knowledge of the averaged density n^* of all pairs with $R \leq R_T$. The stabilization probability (1.9) is not as firmly based as (1.25) and (1.27).

3. TREATMENT WITH DIFFUSIONAL-DRIFT RATE OF COMPLEX FORMATION

Assume upon approach to the reaction R_T-sphere that the energy gained by the (A-B) pairs from their mutual field V(R) is lost upon collision with M so that their relative kinetic energy T_i is restored to their asymptotic equilibrium value E_i. Hence

$$T_i(R) = E_i + V(R+\lambda) - V(R) = E_i + \Delta V, \qquad R \geq R_T \tag{3.1}$$

where λ is the macroscopic mean free path and L_{max}^2 is $(2mT_i)R_T^2$. The microscopic velocity distribution at $R > R_T$ is assumed Maxwellian i.e.,

$$n_i^-(R)dT_i = [(T_i/\pi kT)^{1/2} \exp(-T_i/kT) \, d(T_i/kT)]n(R) \tag{3.2}$$

outside the reaction sphere. The non-equilibrium density n(R) of all AB-pairs with $R \geq R_T$ satisfies the continuity equation $\nabla \cdot J = 0$, which follows from integration of (2.5) over all (E_i, L_i^2). The net outward radial current is

$$\begin{aligned} J &= - \left[D \frac{dn}{dR} + \left(\frac{K}{e}\right) \left(\frac{dV}{dR}\right) n(R) \right] \\ &= - D \exp(-V/kT) \left(d[n(R)\exp(V/kT)]/dR \right) \end{aligned} \tag{3.3}$$

where D and K, the coefficients for relative diffusion and mobility in the gas, are related by De = KkT under equilibrium with the field.

The appropriate distribution for $R \geq R_T$ is therefore (cf. Flannery 1982a,b)

$$n(R_T) = \tilde{n}_A \tilde{n}_B \exp(-V/kT) \left[1 - \left[\frac{\alpha}{\alpha_{TR}}\right] \frac{\tilde{R}}{\tilde{R}_T}\right] \tag{3.4}$$

where the diffusional-drift rate

$$\alpha_{TR} = 4\pi D\tilde{R}_T \tag{3.5}$$

is the rate of formation by transport of pairs with separation $R \leq R_T$ and where, in the presence of interaction V(R),

$$\tilde{R} = \left[\int_R^\infty \exp(KV/De) R^{-2} dR \right] \tag{3.6}$$

replaces R in the customary rate $4\pi DR$ for diffusion alone.

With this analysis, the rate (2.15) yields the standard result (cf. Flannery 1982a,b)

$$\alpha = \frac{\alpha_{RN}\, \alpha_{TR}}{\alpha_{RN} + \alpha_{TR}} \tag{3.7}$$

for termolecular processes in the presence of diffusional drift. The reaction rate therein is identified as

$$\alpha_{RN} = \int_0^\infty dE_i \int_0^{L^2_{max}} dL_i^2 \; v_\infty [1 + \Delta V(R_T)/E_i] \exp[-V(R_T+\lambda)/kT][G(E_i)/L^2_{max}] \times P_i^S(R_T, E_i, L_i^2) \tag{3.8}$$

for general $R_T(E_i, L_i^2)$. As R_T/λ becomes small, as for ion-atom association at most practical gas densities, diffusional-drift can be neglected outside the R_T-sphere and (3.8) reduces to (2.18) with (2.19c) as expected.

Resolution of the dilemma posed in §1 is based on the recognition that (1.28), although similar in form to (3.7) is an approximation only to the reaction rate (3.8) and Expression (1.28) therefore does not contain the physics intrinsic to (3.7). The attractive form of (1.28) tends to mask its approximate character.

4. TERMOLECULAR RECOMBINATION OF ATOMIC IONS AT LOW GAS DENSITIES

Simplifications to the full microscopic treatment as characterized by the appropriate solutions of (2.5) in (2.9b,c) can be therefore achieved by assuming (i) a *reaction radius* R_T - the location R_i of the centrifugal barrier $U_i(L_i)$ for (A^+-B) association or a constant $R_T \sim \frac{1}{2}(e^2/kT)$ for (A^+-B^-) recombination - and (ii) a one-way downward collisional absorption or stabilization rate (2.25) and (2.28) which represents *the strong collision limit*. This rate is given either by the flux across the neck U_i of the effective potential $V_i(R,E_i,L_i^2)$ for A^+-B association, or by the flux across the dissociation limit at $E_i = 0$ for ion-ion recombination since the Coulombic attraction here cannot support an angular momentum barrier.

In order to assess the effectiveness of these two basic assumptions and of other alternative procedures (such as a weak-collision limit and a Variational procedure) which may be adopted, consider the well developed case of termolecular ion-ion recombination at low gas densities. At low N, equilibrium in coordinates R and L_i^2 is quickly established and relaxation in internal energy E_i is the rate limiting step. Integration of (2.5) over (R,L_i^2) yields the standard input-output Master equation

$$dn_i(E_i,t)/dt = \partial n_i(E_i,t)/\partial t = -[n_i(t)\int_{-D}^{\infty} \nu_{if}\, dE_f - \int_{-D}^{\infty} n_f(t)\, \nu_{fi}\, dE_f] \quad (4.1)$$

for the time-dependent pair distribution $n_i(E_i,t)$ per unit energy interval dE_i for bound (E<0) levels in block $\mathcal{E}$ in terms of the frequency ν_{if} per unit interval dE_f for $E_i \rightarrow E_f$ transitions via (AB-M) collisions. This distribution is now expanded as (Flannery 1985)

$$\frac{n_i(E_i,t)}{\tilde{n}_i(E_i)} = P_i^D(E_i)\left[\frac{n_A(t)n_B(t)}{\tilde{n}_A\tilde{n}_B}\right] + P_i^A(E_i)\left[\frac{n(\mathcal{S},t)}{\tilde{n}(\mathcal{S})}\right] \quad (4.2)$$

where $P_i^{A,D}(E_i)$ are the probabilities that E_i-pairs are in the reactive channels which result in association or dissociation respectively. Thus, $P_i^A(E_i \geq 0)$ is zero in the continuum block $\mathcal{C}$ and $P_i^A(-S \geq E_i \geq -D)$ is unity in the block $\mathcal{S}$ of stabilized levels with overall distribution $n(\mathcal{S},t)$. On inserting (4.2) into (4.1) and with the aid of (2.4), the basic expression

$$\alpha\, \tilde{n}_A\tilde{n}_B = \int_{-D}^{\infty} P_i^A\, dE_i \int_{-D}^{\infty} (P_i^A - P_f^A)C_{if} dE_f \quad (4.3)$$

can be derived (Flannery 1985). The one-way equilibrium collision rate for $E_i \rightarrow E_f$ transitions is

$$C_{if}(E_i,E_f) = \int_0^{R_{if}} C_{if}(R)d\underset{\sim}{R} = \int_0^{R_{if}} d\underset{\sim}{R} \int_0^{L_{mi}^2} \tilde{n}_i(R)dL_i^2 \int_0^{L_{mf}^2} \nu_{if}(R)dL_f^2 \quad (4.4)$$

where R_{if} is the lesser of the outermost classical turning points R_i^+ and R_f^+ associated with levels E_i and E_f respectively. For a quasi-steady state (QSS) distribution of pairs in excited block $\mathcal{E}$ $(0 \geq E_i \geq -S)$, (4.3) reduces exactly to the net downward steady-state current (Flannery 1985)

$$\alpha_E\, \tilde{n}_A\tilde{n}_B = -\, j(-E) = \int_{-E}^{\infty} dE_i \int_{-D}^{-E} (P_f^A - P_i^A)C_{if}\, dE_f \quad (4.5)$$

across an arbitrary bound level -E in block $\mathscr{E}$. The probabilities in (4.5) satisfy the QSS-condition

$$P_i^A \int_{-D}^{\infty} C_{if}\, dE_f = \int_{-D}^{\infty} P_i^A\, C_{if}\, dE_f \qquad (4.6)$$

which must rigorously hold in order that α be identified as in (4.5) with a current j(-E) that is constant to variation of E within block $\mathscr{E}$. The equations (4.5) and (4.6) are essentially those as given by the standard QSS-method (Bates and Moffett, 1966; Bates and Flannery, 1968) for recombination under the condition $n_A(t)n_B(t)/\tilde{n}_A\tilde{n}_B >> n(\mathscr{S},t)/\tilde{n}(\mathscr{S})$. The rate (4.5) in practice reduces to a sum of rates each arising from A^+-M and B^--M collisions, respectively (Bates and Flannery, 1968; Flannery and Yang, 1980).

4.1. One-Way Equilibrium Rate: *Strong Collision Limit*

On ignoring upward transitions across E = 0 for $R \leq R_T$ and upon assuming full thermodynamic equilibrium (which entails equal upward and downward collisional rates at each R) for $R \geq R_T$ then (4.5) provides the one-way downward equilibrium flux across E = 0 as

$$\alpha_T(R_T)\ \tilde{n}_A\tilde{n}_B = \int_0^{\infty} dE_i \int_0^{R_T} d\underset{\sim}{R} \int_{-D}^{0} C_{if}(R)dE_f \qquad (4.7)$$

which is the modern collisional equivalent of the low density treatment of Thomson (1924). Figure 2 displays the variation obtained by Flannery and Mansky (1987) of the ratio (α_T/α_E) with R_T for equal mass species A^+, B^- and M. Energy-changing collisions via various (A^+-M) interactions CX for symmetrical resonance charge transfer (Flannery 1980), HS for hard sphere encounters (Flannery 1981) and for polarization collisions (Bates and Mendaš 1982) are considered. Equivalence between α_E and α_T occurs for $R_T \sim 0.5\ (e^2/kT)$, a value that can be assigned only after detailed calculations for α_E. Table 1 displays the assigned R_T for various values of the mass ratio parameters

$$a = M_B M/M_A(M_A+M_B+M) \qquad (4.8)$$

where $M_{A,B}$ and M are the masses of the ions and gas atoms respectively. The very small values of R_T for small a $\sim 10^{-3}$ are simply a manifestation that (4.7) becomes invalid for electron-ion recombination since then only a small fraction $\delta = (2m_e/M)$ of the electron's energy can be transferred in e-M collisions so that the upper limit of the E_i-integration in (4.7) is $\delta e^2/R_T$ rather than infinity (Bates, 1980).

The divergence of (4.7) illustrated in Fig. 2 as $R_T \rightarrow \infty$ results from

Table 1: Radius R_T (in units of e^2/kT) so assigned that the one-way equilibrium rate (4.7) reproduces the exact rate (4.5) for various mass parameters (4.8) and (A^+-M) interactions.

a	CX	HS	POL
0.001	0.1200	0.1655	0.0922
0.010	0.2277	0.2902	0.1801
0.100	0.3657	0.4510	0.3074
0.333	0.3992	0.5076	0.4861
1.000	0.5449	0.5187	0.3749
10.000	-	0.4211	0.4280
100.000	-	0.2444	0.2735
1000.000	-	0.1188	0.0922

the divergence in the equilibrium density $\tilde{n}(E_i)$ of Coulombic levels as $E \to 0$, and from the neglect in (4.7) of upward collisional transitions within the R_T-sphere, an assumption which is inappropriate as $R_T \to \infty$.

This divergence can be eliminated, not only by maintaining R_T finite, but also by considering the one-way equilibrium rate

$$\alpha_B(-E)\ \tilde{n}_A\tilde{n}_B = \int_{-E}^{\infty} dE_i \int_{-D}^{-E} dE_f \int_0^{R_{if}} C_{if}(R)d\underset{\sim}{R} \tag{4.9}$$

across any bound level -E in block $\mathcal{E}$. The variation (Fig. 3) with -E of this rate displays a minimum (Flannery 1980, 1981) at the bottleneck energy $E^* \underset{\sim}{\sim} 2kT$. Thus $\alpha_B(-E^*)$ is the least upper limit. This variational procedure is akin to the Wigner-Keck Variational phase-space treatment (Wigner, 1937; Keck, 1967). Table 2 provides (α_B/α_E) for various mass parameters a. The procedure is more reliable for a in the range $0.1 \leq a \leq 10$ where the collisional dynamics for larger energy-transfers are important.

4.2. Diffusion Method: *Weak-Collision Limit*

For small energy-transfers, the collision integral in (2.5) when integrated over R and L_i^2 can be represented in differential form, so that the diffusional net current in energy space is (cf. Flannery 1987)

$$j_D(E_i) = -\ D_i^2(E_i)\ (\partial P_i^D/\partial E_i) \tag{4.10}$$

where the energy diffusion coefficient is

$$D_i^2(E_i) = \frac{1}{2}\int_{-D}^{\infty} (E_f-E_i)^2\ C_{if}\ dE_f \tag{4.11}$$

Solution of (4.10) subject to steady state current j_D within block $\mathcal{E}$ and to the conditions to P_i^A at the boundaries to block $\mathcal{E}$ yield

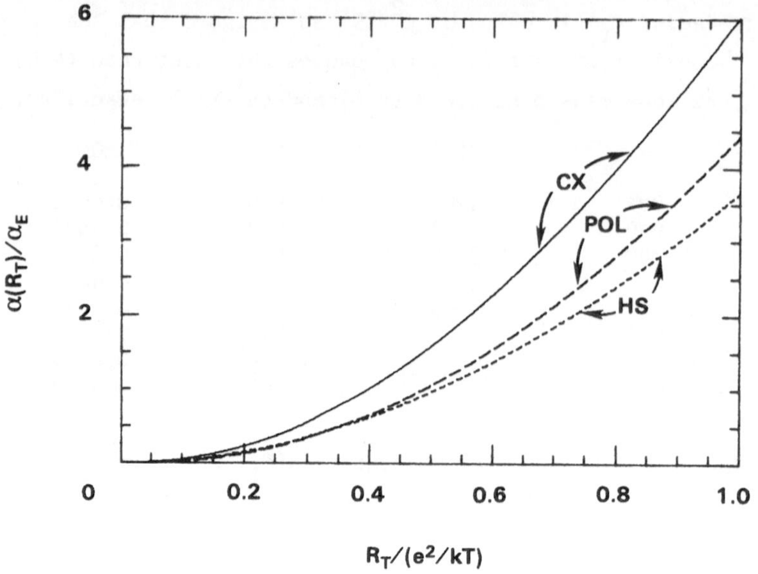

Fig. 2. *Strong Collision Limit:* Variation with R_T of the one-way equilibrium rate, (4.7) of text, across energy level E = 0.

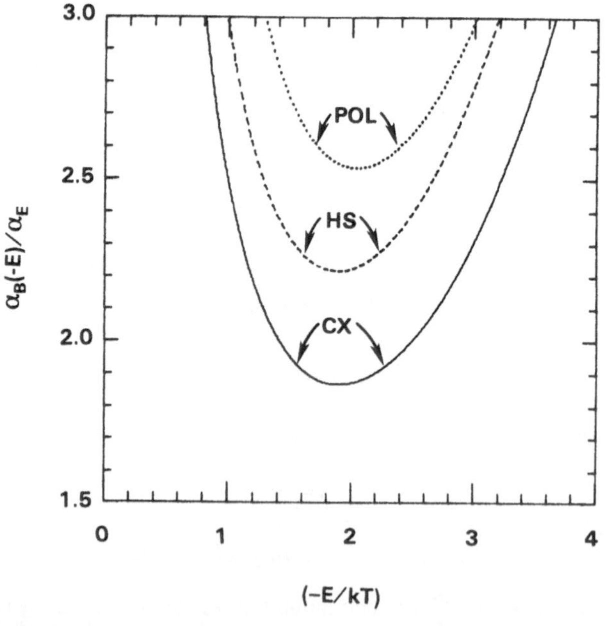

Fig. 3. *Bottleneck Method:* One-way equilibrium rates, (4.9) of text across excited levels with energy -E relative to the dissociation threshold.

Table 2: Ratio of α_P/α_E and α_B/α_E from (4.13), (4.9) and (4.5) for various mass ratios (4.8) and (A^+-M) interactions.

	α_P/α_E			α_B/α_E		
a	CX	HS	POL	CX	HS	POL
0.001	1.0275	1.0127	1.0008	16.9958	25.7816	32.4466
0.010	1.3040	1.2216	1.1634	5.5130	7.3362	8.3690
0.100	3.5223	2.7388	2.1306	2.3835	2.9386	3.3538
0.333	6.8404	4.9678	3.3611	1.8645	2.2155	2.5414
1.000	9.2715	6.6036	4.0598	1.7222	2.0152	2.3328
10.000	------	3.5103	2.1306	----	2.7464	3.3537
100.000	------	1.4550	1.1634	----	6.3015	8.3690
1000.000	------	1.0934	1.0008	----	20.2334	32.4466

$$P_D^A(E_i) = \left[\int_{E_i}^{0} dE/D_i^{(2)}(E)\right]\left[\int_{-S}^{0} dE/D_i^{(2)}(E)\right] \qquad (4.12)$$

so that the steady-state downward diffusional current across E = 0 is then

$$-j_D(0) = \alpha_P\,\tilde{n}_A\tilde{n}_B = \left[\int_{-S}^{0} dE/D_i^{(2)}(E)\right]^{-1} \qquad (4.13)$$

This is the original result of Pitaevskii (1962) who designed it only for electron-ion recombination in a gas. Table 2 shows that it is indeed correct in this collisional limit of large a. Its departure from α_E for general mass ratios arises from identification in (4.13) of α with a current based on (4.10) or (4.12), which does not satisfy the exact QSS-condition (4.6). When (4.12) is inserted in the basic rate (4.3) to give α_D, excellent agreement is apparent (Fig.4) for all mass ratios a and A^+-M interactions. The rates in Fig. 4 are normalized (Flannery 1980, 1981, 1987) to Thomson rates. In the limit of small energy transfers (small and large a) the current (4.13) is so small and agrees essentially (as in Table 2) with the result of (4.3).

4.3. *Variational Principle*

It has recently been proposed (Flannery 1987) that termolecular association proceeds in such a manner that the overall association rate (2.4) is a minimum at any given time. Under the separation (4.2) which is necessary for emergence of the QSS condition (4.6), a Variational Treatment of the QSS approximation can therefore be constructed by varying P_i in (4.5) so as to yield minimum rates α. With $\lambda = -E_i/kT$, the simplest one-parameter variational form

$$P_V^A(\lambda;\lambda_*) = 1 - (1+\chi)\exp(-\chi) \qquad \chi = \lambda/\lambda_* \qquad (4.14)$$

which tends to zero as $E_i \to 0$ and to unity as $E_i \to -\infty$ (rather than to unity as $E_i \to -S$), yields the exact QSS rate α_E of (4.5) for $\lambda^* \simeq 1.2$. Fig. 5 illustrates the comparison between $P_V^{A,D}$, $P_D^{A,D}$ and $P_E^{A,D}$ which are respectively the variational diffusional, and exact QSS probabilities for

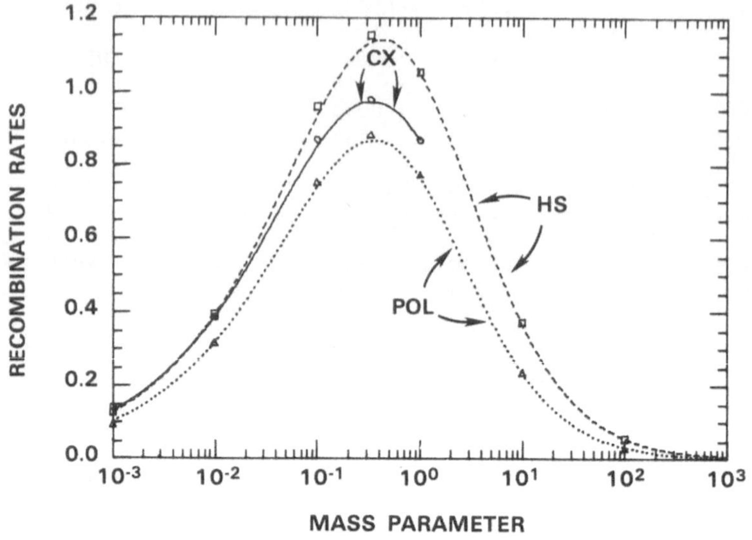

Fig. 4. Variation of QSS-rates (lines) and diffusion rates (symbols) with mass parameter a.

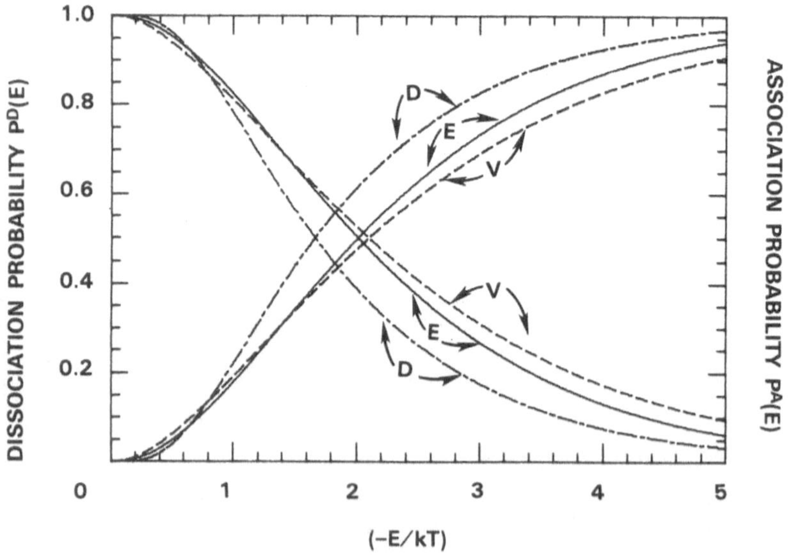

Fig. 5. Probabilities for dissociation and association of pairs with internal energy -E relative to the dissociation limit.

termolecular recombination of equal mass species under a polarization A^+-M attraction. More elaborate expressions with two and three variational parameters-yield probabilities closer to the exact $P_E^{A,D}$ (Flannery 1987). The overall rate α is however not sensitive to these changes as suggested by the close agreement between the diffusion and exact results of Figs. 4 and 5. All of the adopted variational expressions produces α_E exactly.

Acknowledgement: On this occasion of the seventieth birthday of Sir David, I would like to dedicate this work to him. It is but only a small token relative to the valuable contributions and many stimulating lines of enquiry Sir David has provided and continues to provide to this field. I would also like to thank Dr. E. J. Mansky of this School for help in preparation of the Tables and Figures. I also thank Audrey Ralston for her pioneering efforts in adapting the latest computer and laser technology to the overall production of the paper. This research is supported by the US Air Force Office of Scientific Research under Grant No. AFOSR-84-0233.

References

Bass, L., Chesnavich, W. J., and Bowers, M. T., 1979, J. Amer. Chem Soc., 103:5493
Bates, D. R., 1978, Proc. Roy. Soc. Lond. A., 360:1
Bates, D. R., 1979a, J. Phys. B. Atom. Molec. Phys., 12:4135
Bates, D. R., 1979b, J. Chem. Phys., 71:2318
Bates, D. R., 1980, J. Phys. B. Atom. Molec. Phys., 13:2587
Bates, D. R., 1986, J. Chem. Phys., 84:6233
Bates, D. R., 1985, J. Chem. Phys., 83:4448
Bates, D. R., and Flannery, M. R., 1968, Proc. Roy. Soc. Lond. A., 302:367
Bates, D. R., and Mendaš, 1982, J. Phys. B. Atom. Molec. Phys., 15:1949
Bates, D. R., and Moffett, R. J., 1966, Proc. Roy. Soc. Lond. A., 291:1
Blades, A. T., and Kebarle, P., 1983, J. Chem. Phys., 78:783
Flannery, M.R., 1980, J. Phys. B. Atom. Molec. Phys., 13:3649
Flannery, M.R., 1981, J. Phys. B. Atom. Molec. Phys., 14:915
Flannery, M. R., 1982a, Ion-Ion Recombination in High Pressure Plasmas, in: "Applied Atomic Collision Physics, Vol. 3", E. W. McDaniel and W. L. Nighan, eds., Academic Press, New York.
Flannery, M. R., 1982b, Phil. Trans. Roy. Soc. Lond. A., 304:447
Flannery, M. R., 1985, J. Phys. B. Atom. Molec. Phys., 18:L839
Flannery, M. R., 1986, J. Phys. B. Atom. Molec. Phys., 19:L227
Flannery, M. R., 1987 (in preparation).
Flannery, M. R., and Mansky, E. J., 1987 (in preparation).
Flannery, M. R., and Yang, T. P., 1980, J. Chem. Phys. 73:3239.
Forst, W., 1973, Theory of Unimolecular Reactions, Academic Press, New York.
Herbst, E., 1979, J. Chem. Phys., 70:2201
Herbst, E., 1980, J. Chem. Phys., 72:5284
Keck, J. C., 1967, Adv. Chem. Phys., 13:85
Klots, C. E., 1971, J. Phys. Chem., 75:1526
Loeb, L. B., 1955, "Basic Processes of Gaseous Electronics", Ch. 6, Univ. of California, Berkeley.
Mahan, B. H., and Person, J. C., 1964, J. Chem. Phys., 40:392
Marcus, R. A., 1975, J. Chem. Phys., 62:1372
Pitaevskii, L. P., 1962, Soviet Physics-JETP 15:919
Thomson, J. J., 1924, Phil. Mag., 47:337
Troe, J., 1977, J. Chem. Phys., 66:4758
van Koppen, P. A. M., Jarrold, M. F., Bowers, M. T., Bass, L. M., and Jennings, K. R., 1984, J. Chem. Phys., 81:288
Viggiano, A. A., 1986, J. Chem. Phys., 84:244
Whitten, G. Z., and Rabinovitch, B. S., 1963, J. Chem. Phys., 38:2466
Wigner, E., 1937, J. Chem. Phys., 5:720

AN EXPERT DATABASE ON ATOMIC AND MOLECULAR PHYSICS

F.J. Smith

Department of Computer Science
School of Physics and Mathematical Sciences
The Queen's University of Belfast
Belfast BT7 1NN United Kingdom

ABSTRACT

A database of numerical data on atomic and molecular physics relevant to fusion has been built with the emphasis on reliability and accuracy rather than on quantity. Since leading experts in any scientific discipline are always the best people to recommend data in that discipline, we have used leading Atomic Physicists, both experimental and theoretical, to decide each and every numerical value in the database. The School of Physics and Mathematical Sciences at Queen's, set up by Professor Sir David Bates, was an ideal location for the database as it provided the expert experimentalists, the expert theoreticians and, in addition, the Computer Scientists needed to design and manage the database system.

INTRODUCTION

The aim of this talk is to describe the database system at Queen's on atomic and molecular physics relevant to fusion; the emphasis of the talk will be on the database system and its management rather than on the data itself or on its use. The system is supported by the United Kingdom Atomic Energy Authority, Culham Laboratory, to provide data needed mainly by physicists studying the magnetic confinement of a plasma within a Tokamak. However, the data also have obvious applications to inertia fusion, astrophysics and solar physics.

The database system is a project of the School of Physics and Mathematical Sciences. This School was set up by Professor Sir David Bates twenty years ago (he remained its chairman until he retired in 1982), and consists of four departments, three of which contribute to the database system: the Department of Applied Mathematics and Theoretical Physics and the Department of Pure and Applied Physics provide the expertise needed to authenticate and assess both experimental and theoretical data and to recommend the best data available and the Department of Computer Science provides the database system and the management of the database. It is this combination of Computer Scientists, interested in scientific database systems, and Physicists, both theoretical and experimental, which we believe has provided the ideal background for our database system.

The role of the theoretical and experimental physicists is not unexpected in the building of such a database system. However, the involvement of

Computer Science is more unusual and arises out of the development of computat Queen's. I will therefore digress for a few moments to discuss the development of computing and Computer Science at the University; they are intimately tied up with the research work of Professor Sir David Bates and his colleagues at Queen's over the last fifty years.

COMPUTING AT QUEEN'S

David Bates' interest in computing began with the computing device shown in Figure 1. This is a form of mechanical analog computer, called a differential analyser, which was used by David Bates in his early research work on the upper atmosphere at Queen's in the 1930's as a student for an MSc (1) under Dr Massey (later to become Sir Harry Massey at University College, London). This machine has a history linked to Queen's which goes back a hundred and twenty years, as it was based on an invention by James Thompson who was the second Professor of Civil Engineering Science at Queen's College, Belfast, from 1857 to 1873. While at Queen's he invented a mechanical analog calculator or integrator between the years 1861 and 1864 - this integrator could compute the area under a curve (2). It consisted essentially of a moving cylinder, sphere and disc and in the diagram there are four such integrators of similar design linked together. Since the integrator was first used in a paper to the Royal Society by James' brother William, better known as Lord Kelvin (3), the literature often mistakenly attributes the invention of the device to Kelvin. Kelvin used his brother's integrator as a harmonic analyser to predict heights of tides in ports round the British Empire. He pointed out that his brother's integrators could be linked together to solve coupled differential equations, but it was not until this century that this idea was put into practice in a machine which was called a differential analyser. One of these built in Manchester was by Hartree and Porter in 1935 (4). This stimulated interest in the idea in Belfast and with advice from Professor Hartree the machine you see in the figure was then built by Mr Wylie and Mr Buick in the Physics Department for Dr Massey and his assistant Dr R.A. Buckingham (now Professor Emeritus of Computer Science of Birkbeck College, London) in the mid-1930's (its cost being less than £50) and reported in detail in the paper given to the Royal Irish Academy in 1938 (5). This computer lies at the beginning of our story because, as I have said, it was used by David Bates when he was a student and introduced him to the idea of automatic computation which was to have a fundamental effect on the development of computing at Queen's in later years.

Dr Massey's analog computer went to London during the war and was unfortunately destroyed by a bomb there, in the basement of University College. David Bates also went to England during the war where he worked with Dr Massey on de-Gaussing ships, to help them avoid mines. After the war he went to University College, London, but returned to Queen's in 1951 as Professor of Applied Mathematics where he continued the tradition of a computational approach to the understanding of Atomic Physics which he began with the differential analyser in Figure 1.

However, this early form of analog computer was inaccurate; so when David returned to Belfast, calculations of cross sections, phase shifts, wave function, etc. depended on numerical computations using mechanical calculators (such as the Brunswiga) or later electrical calculators (such as the Facit). These were slow and laborious: research students would have to slave for months, putting in hours of manual computation each day to calculate a set of energy levels or phase shifts to appear in one of those early Atomic Physics papers coming from Belfast. For example the seminal paper by Bates, Ledsham and Stewart in 1953 (6) on the wave functions for H_2^+ was carried out on a Brunswiga and required three man/woman years of hard computing.

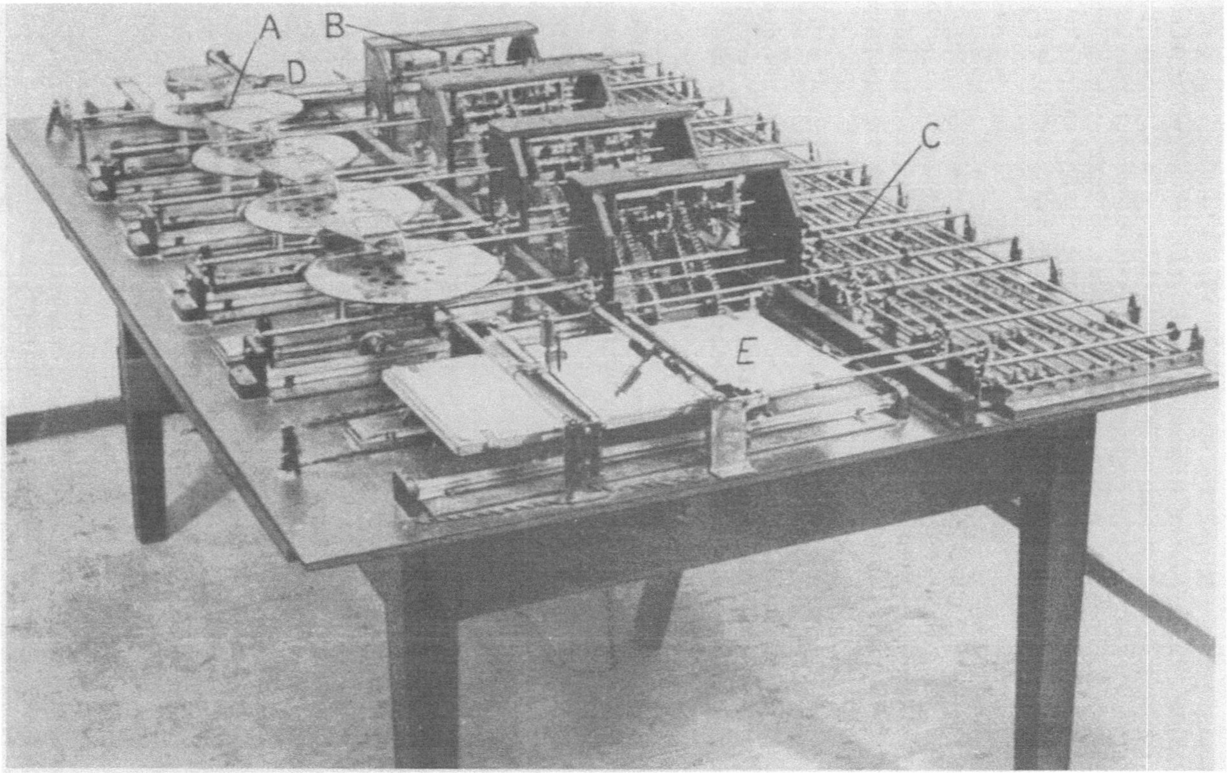

Figure 1

Differential analyser, a mechanical analog computer, used by David Bates when a student of Dr Massey at Queen's University, Belfast, in the late 1930's.

This labour was tedious, frustrating and holding back research; so David was quick to appreciate the importance of digital computers, their enormous speed relative to the hand calculators and their greater accuracy relative to analog computers when they first became available in the late 1950's. When Short Brothers and Harland Ltd., the aircraft company in Belfast, purchased an English Electric DEUCE computer in 1958, David arranged for his Department of Applied Mathematics (and indeed any other user at the University) to get access to the computer in the evenings. The enormous increase in computing power opened up new possibilities and for example, it enabled Burke, Burke, Percival and McCarroll, 1962 (7) to carry out the first realistic electron-hydrogen close coupling calculation. In Figure 2 I show a photograph of one of these early English Electric DEUCE computers similar to the one in Shorts. Many of us have recollections of late nights, the noise of punched cards, mechanical tabulators and sorters and the sight of dawn appearing behind the cranes of the shipyard as we left in the early morning.

With the advent of the new computer at Shorts a lectureship in Digital Computing was created within the Department of Applied Mathematics in 1958 - one of the first lectureships in Digital Computing in the United Kingdom and the post was taken by Dr McCarroll, now Professor of Physics in the University of Bordeaux.

The use of the DEUCE computer at Shorts grew rapidly and a few years after Dr McCarroll's appointment, in 1961, the University bought its own English Electric DEUCE computer from a generous grant from the United States Navy, and Dr Alex Dalgarno (now Professor of Astronomy at Harvard) was appointed Director of the new Computer Laboratory.

This DEUCE computer had a computer power not any greater than that of a personal computer which most of us have on our desks nowadays, and since it had to be programmed with binary instructions it was one or two orders of magnitude more difficult to use. But for its sole purpose, scientific calculations, it was an enormous advance on the manual calculators we depended on before this time. Also, it was well suited to those calculations because of its software; programmes for numerical computation were written for DEUCE by some of the best scientific minds in the country at the time and were freely exchanged between all computational scientists. Its biggest problem was the slow means of input and output (punched cards) and it broke down frequently. It is interesting to note that the Queen's DEUCE computer needed two full-time engineers to keep it running and it rarely lasted a whole day without a fault. Compare that with your PC which typically runs for months (or even a year or two) without failure.

In the meantime, in the early 1960's the development in the use of computers in Atomic Physics grew rapidly at Queen's and within two years outgrew the DEUCE; soon Queen's became one of the largest university users of computers in the United Kingdom. Amongst the computers used were the IBM 7090 (using FORTRAN) at Imperial College, London, the Atlas Computer (using Atlas Autocode) in Manchester, the Atlas Laboratory at Harwell (using FORTRAN and Algol), the Elliott 503 in Bristol (using Algol) and more computers in the United States including some NASA computers when not being used on space ventures, and the IBM 7090 at the University of Maryland. Each year, in the early and mid-1960's, many of the staff of David Bates' Department of Applied Mathematics spent the summer months in the United States to get access to the large computational capacity there (having exhausted all of the capacity available in the United Kingdom).

Figure 2

Photograph of a DEUCE computer similar to the first digital computer used by David Bates' Department of Applied Mathematics from 1961 to 1965. It used valves.

Seeking larger and more powerful computers was not the only means of the pursuit of faster and more accurate computation in atomic physics. The study of numerical analysis also made a major contribution and at the same time became the seed round which formed the Department of Computer Science.

In 1965 the University appointed its first Professor of Computer Science, Professor Jim Browne, now Head of the Department of Computer Science at Austin in Texas. The University also purchased in the same year a much more powerful computer than the old DEUCE, the ICL 1907, which was based on solid state technology, i.e. on transistors rather than on valves. It was much more powerful than DEUCE, but more important, it was many times more reliable and it could be programmed in Fortran or Algol, languages with which we were already well familiar on other computers. A photograph is shown in Figure 3 and when it was delivered it had some disc handlers each of which enabled us to record up to four million characters of data at a time on a set of discs - this seemed an enormous amount of data to us at the time. But a few years later we took delivery of an even larger fixed (Bryant) disk, with 100 Mbytes of storage. So, for a University of our size, and indeed for any University, we had excellent computing facilities, thanks to David Bates; we had access to large computers in England and in the United States and we also had excellent computing facilities on site. These included this large disc storage and we had to think again how we might use such massive storage. I can remember Professor Jim Browne suggesting to me, about 1967, that it might be a useful area of research to investigate how we might use such large storage. Out of that idea came the development of an on-line database in atomic physics a few years later.

In October 1968 the Computer Science Department became a separate department from Applied Mathematics, but still tied to Applied Mathematics as part of a new School of Physics and Mathematical Sciences which was set up in 1964 to join closer together the Departments of Physics and Applied Mathematics and later also Pure Mathematics. David Bates chaired the new School and continued his chairmanship until his retirement in 1982.

Just before Computer Science became a separate department, as I have stated, the database on atomic physics had already been started within the Department of Applied Mathematics. Once the break occurred, it was natural for the new Department of Computer Science to take this project with it and continue this (and a few other) links with atomic physics. This gave our database two important advantages: close association with the subject (i.e. atomic physics) and relatively impartial and independent management.

EARLY DATABASE WORK

The first work on atomic data was in early 1968, and it began with the collection of data on interatomic potentials; these could then be used to compute many of the macroscopic properties of atoms and molecules, such as the transport properties of atomic or molecular gases. Work was supported by the new Office of Scientific and Technical Information (set up in the United Kingdom by the new Government of Harold Wilson in the late 60's) which was more interested in the technology of on-line databases than in the data. With substantial and generous support from OSTI an on-line database system for the retrieval of interatomic potentials was built by 1970, possibly the first on-line numerical database system in Europe, and it was demonstrated at the Amsterdam ICPEAC Conference in 1971, through a terminal linked over a telephone line to the ICL 1907 computer in Belfast. Although technically the on-line data system was successful, indeed very successful, in achieving its technical aim of retrieving data quickly on a terminal over a telephone network (8), it was not financially viable in such a limited field as atomic and molecular physics: there were just not enough

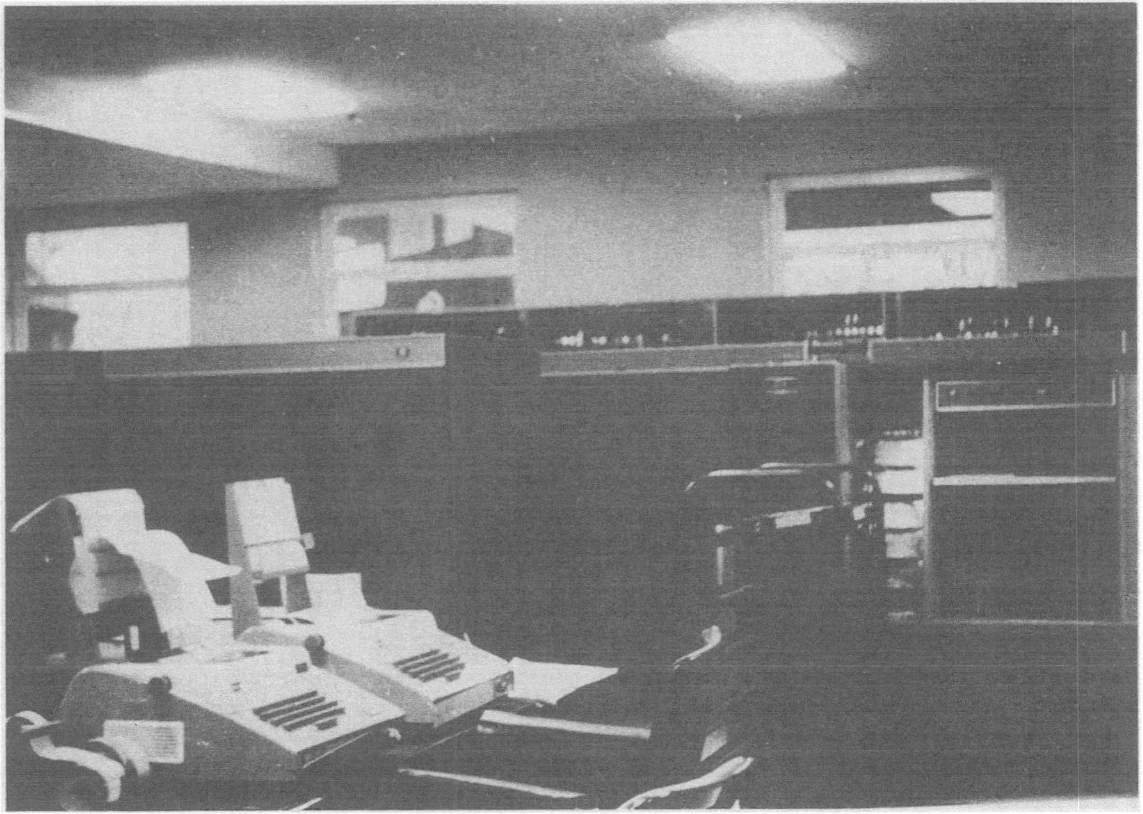

Figure 3

Photograph of the ICL 1907 computer (about 1 MIP) purchased for David Bates' Department of Applied Mathematics in 1965. Unlike its predecessor, it used transistors and was programmed in FORTRAN and ALGOL and later, in 1971, it was the second computer to be programmable in PASCAL.

scientists in the world with a terminal and modem on their desk to make an on-line service even remotely viable. Even today, an on-line service in atomic physics is not viable on its own; however, it is interesting to note that this system researched and demonstrated a technology which is at the basis of many on-line financial and on-line engineering databases available today (9).

Because of lack of users our funding for these new technologies was inevitably lost. However, the University fortunately maintained an interest in atomic data and when the Princeton Tokamak recorded a temperature of 60 million degrees, after they changed from a steel to a carbon delimiter within their Tokamak (10), Queen's was well positioned to become involved in the search for better data to describe what was happening within the plasma. The plasma modellers had at last realised that without an understanding of atomic and molecular physics, they would be unable to understand fully what was happening within their thermonuclear plasmas.

The International Atomic Energy Agency in Vienna was given the task of starting an international program to collect atomic and molecular data for fusion and following a meeting of plasma physicists and atomic physicists at Culham in 1976 (11), Queen's University became the main United Kingdom Centre for data collection, with an emphasis on authenticated or recommended numerical data. Support was given by the Culham Laboratory and by Euratom and has continued until today. A new contract until 1989 has just been agreed.

When we started to collect data again in 1977, this time atomic and molecular data relevant to fusion rather than interatomic potentials, we were able to avoid one of the main mistakes we had made in the early years between 1967 and 1975: the use of a technology well beyond the needs and wishes of our users. Our earlier system had collected a great deal of data on interatomic potentials; but unfortunately it was never used. First of all, as we have said, it could only be accessed through a terminal and scientists were then not al all accustomed to accessing databases with terminals (indeed even today they are still not accustomed to it). But further to this, we were collecting data in an area where there was not a great demand. The area of interatomic potential energies was not a growing area in atomic and molecular physics at that time (it might have been more successful a decade earlier); so we had few people interested i.e. we were collecting data which was not relevant to the current needs of the bulk of atomic physicists. This was our second error; but there was a third. Although we collected data, we did not do anything to add to its value. The staff in the Applied Mathematics Department and then later the Computer Science Department who collected the data were young graduates or assistants who did not have the authority nor the experience to make their data acceptable to the main body of physicists who might have been interested in them. Only well known experts in any field will be accepted as reliable providers of data by the bulk of scientists within that field. If anyone other than experts provide the data, for very good reasons, it is likely to be viewed with suspicion and not used.

We set out to correct these three errors in the new database we began to build on atomic and molecular physics of fusion.

THE NEW DATABASE SYSTEM

The fundamental principle on which our new database system was built was that it should be user driven rather than designer driven; that is, at all times we would find out from the user body what they actually needed, rather than what we, as database providers, believed they needed and wanted. This was the common feature of the three faults that we had made in our early system - and indeed it is still this lack of knowledge of real users' needs which holds back today most database systems. (It is interesting here to note that many books on database systems on the market today do not even mention the users or the users' needs. Is it surprising that systems are generally so user unfriendly?).

However, to be truthful, we had little choice but to make our database user orientated because our funding came from the United Kingdom Atomic Energy Authority, Culham Laboratory, who were to be the main British users of our database and they made it clear from the beginning what was needed, particularly the data which was most needed, and our funding depended on it.

As we were a member of a network of computer centres set up under the auspices of the International Atomic Energy Agency to collect atomic and molecular data relevant to fusion, we were also able to ensure that what we were doing complemented, rather than overlapped with work by our colleagues in other countries. This international co-operation helped us: for example the International Atomic Energy Agency was collecting a bibliographic index to data; so we did not have to do any literature searching. We were able from the beginning to depend on the Agency's publications, particularly the Bulletin on Atomic and Molecular Data for Fusion which provided a quarterly index to relevant data in the literature (12) and the retrospective Computer Index on Atomic and Molecular Data for Fusion (CIAMDA) (13).

With this help we were able to begin immediately with the collection of numerical data, the data most needed, rather than bibliographic data, making sure that it complemented rather than overlapped with what others were doing. After consultation, it was agreed with Culham that we would collect data on electron ionization of atoms; this was supported by the IAEA as it did not overlap with the work of the other data centres. We began with the light atoms and ions, from hydrogen to oxygen, which were the most important and appeared in our first report (14) and later worked on heavier atoms and ions, the subject of our second publication (15).

EXPERTS

We wanted to ensure that the data had the highest possible quality. This required that it must be recommended by experts, preferably by at least one experimentalist and one theoretician, but not be someone who was directly involved in the measurements or calculations. (Scientists cannot be asked to comment impartially on their own work!). The problem was how to acquire the expert's knowledge; how does one get the best experts in any field to collect data? A leading scientist does not want to give up much of his time for such a routine activity as collecting old data already published in the literature. Obviously an active scientist wants to spend his time researching new ideas, not collecting old ones.

Our approach was basically to keep the total amount of time which had to be spent by each expert to an absolute minimum. The expert had only to advise, make decisions on which data were best, check what was done, resolve the difficult problems. To help we had to employ full time staff (usually people with a PhD in atomic physics) who did <u>all</u> of the routine work.

Let us describe what we did and illustrate with particular attention to one reaction, the electron ionization of C^{3+}.

Figure 4 shows diagrammatically our procedure. A file would be prepared on each reaction and photocopies of any relevant publications included. An appointment for a short consultation with each expert in his office would be made and he would be asked about several reactions. Usually in a matter of a minute or two the expert could select from the reprints for each

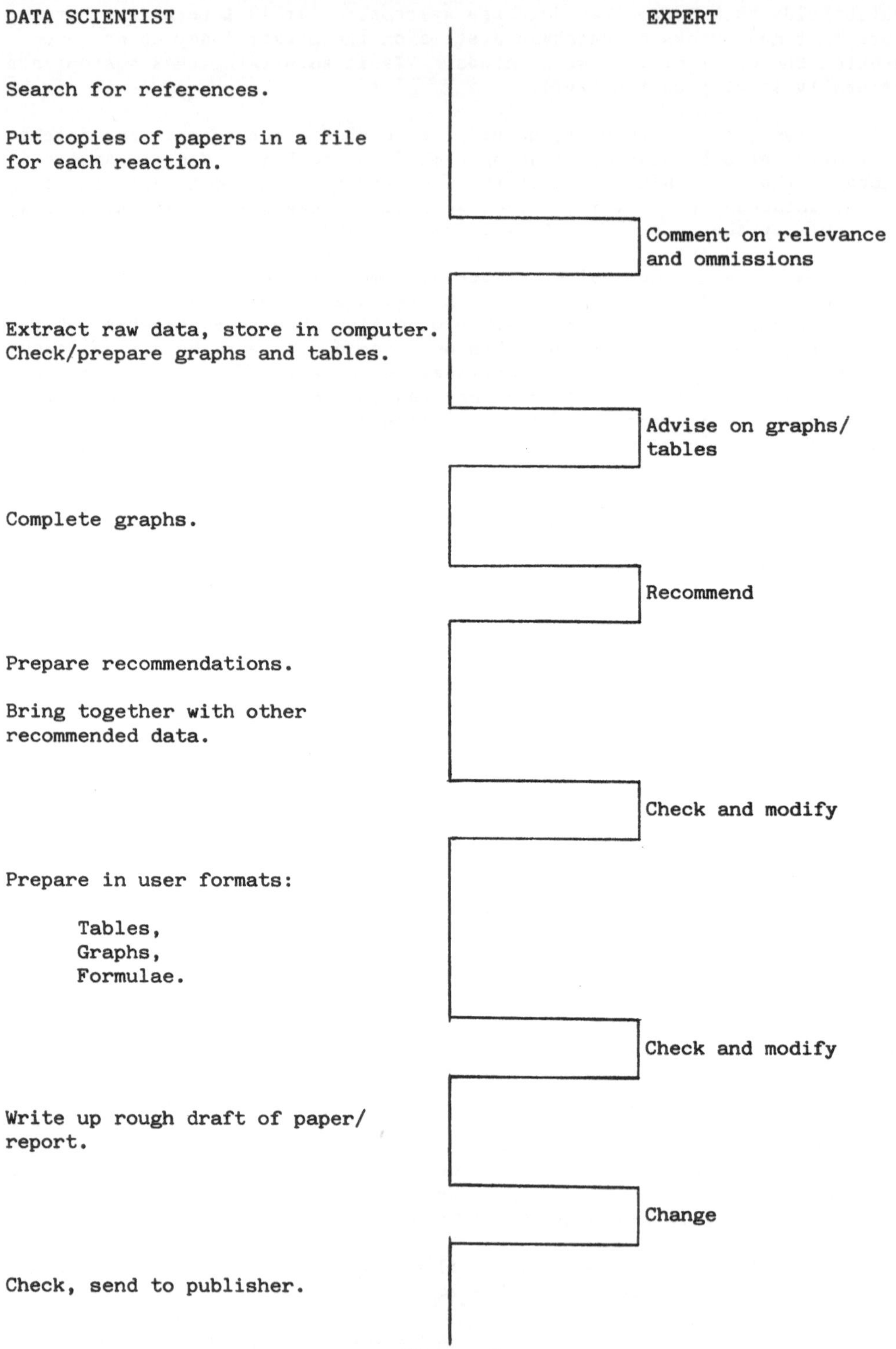

Figure 4

Illustration of our procedure for the interrogation of experts, minimizing the time of the experts.

reaction those publications which were worth further processing - often only 1 or 2 , sometimes as many as 10, all too often none! Sometimes the expert would advise on a paper missed and this would have to be photocopied later and brought back at a later consultation. Sometimes the expert could advise on some recent measurement not yet in the literature.

The recommendations of two or more experts obtained in this way would be put together, the data from the publications (if any) would then be extracted from the literature and displayed together, on a graph. For example, the data from 4 sources on the electron ionization of C^{3+} are shown in Figure 5. Then the file, now containing in this case the four original papers and the four curves together on one graph paper (along with the Lotz empirical curve), would be brought, with other similar files, to the experts and the experts asked which of the four data sets was likely to be most accurate and which should be rejected in each region of the graph. The experts would usually recommend one particular experimental data set and possibly some theoretical form for those higher (or lower) energies not covered by experiment. In some cases one particular set of results would be recommended for low energies, and another set for high energies and a smooth curve would have to be drawn between these. By such means, a recommended curve was determined on the advice of the experts. This was drawn onto the graph and in a later consultation the recommended data was finally checked and recommendations on its accuracy were again determined by the experts.

As we have explained, since the experts are conversant with the field they normally only require a few minutes or less to advise on the recommended curve for each reaction; so the total amount of time taken up is minimised. If the experts should be uncertain they may ask the database staff to perform some further calculations or find some other paper or some other relevant information. As before, the experts require little time to do this; all of the routine work is done by the database staff. The extra information is brought back and again the experts' judgement is sought.

DISAGREEMENTS

Usually we have found that the experts agree. But in cases where they did not agree, further consultation usually results in a change of recommendation by one or other of the experts. Only in one case, the example we have chosen, C^{3+}, did disagreement need a meeting of all experts to resolve. The decision in our first report [14] was to recommend the theoretical calculations of Jacubowitz rather than the experimental data of Crandall. Later measurements showed that Crandall were correct and our recommendation had to be changed [15].

RECOMMENDATIONS WHERE THERE ARE NO DATA

All of the above procedures and consultations are fairly straightforward when there are a number of measurements or calculations for a particular reaction. However, often there are no data, or the data that are available are known to be highly unreliable. Then some expirical rules may be used to obtain the best recommended data. Again our experts will advise us on which rules to use. For example, for our database on electron ionisation, classical scaling has been used as illustrated in Figure 6. This shows that a comparison of electron ionisation cross-sections for an iso-electronic sequence demonstrates that the scaled data are not far apart (except when autoionization makes a major contribution as in the case of 4 ions in Figure 6); so we are able to use scaled data in cases where no experimental or theoretical data exists, for example we used scaled data for Si^{11+}, but with larger error bars.

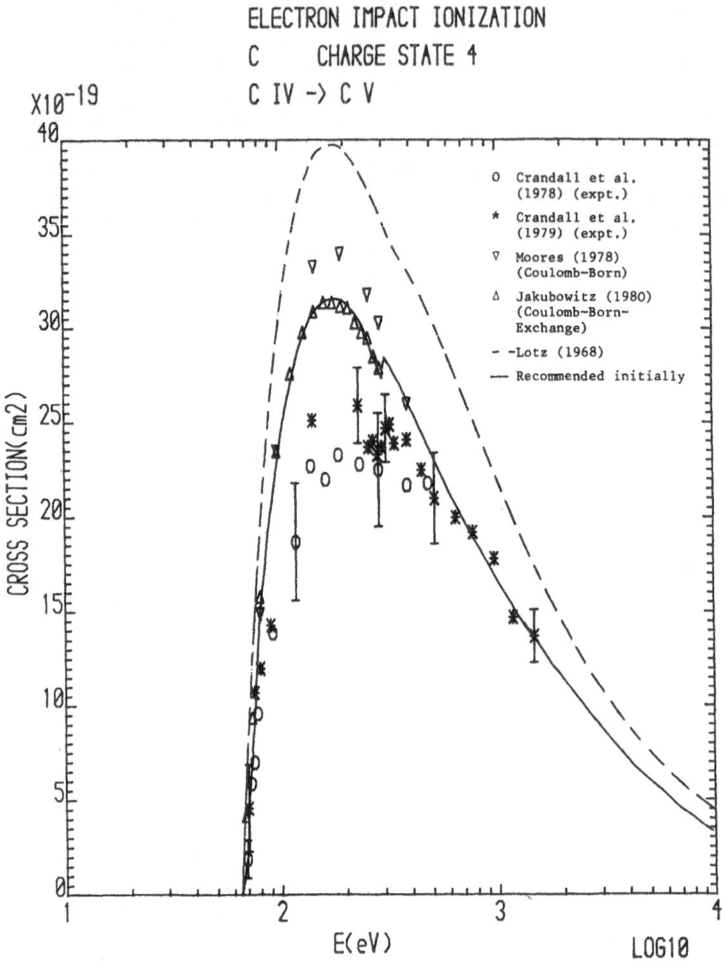

Figure 5

Data from several sources on the electron ionization of C^{3+}. Graphs such as this would be shown to our experts and they would be asked which data to recommend at each energy, ie to recommend a curve giving "recommended data".

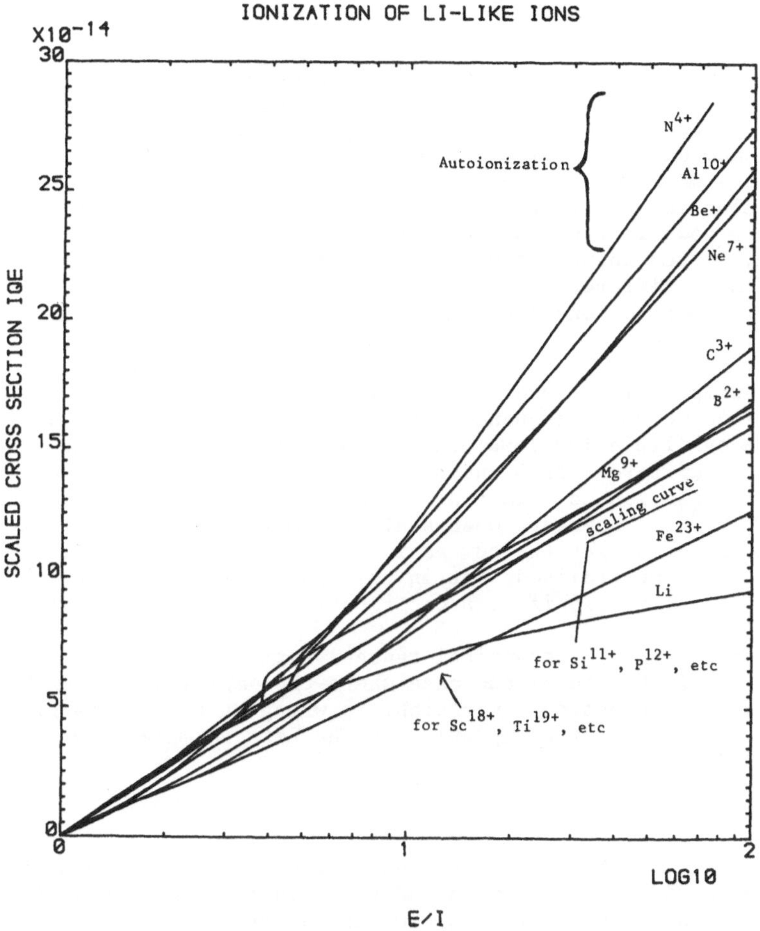

Figure 6

Classically scaled data, the electron ionization of the Li iso-electronic sequence. If the classical scaling law was obeyed exactly by all ions in the sequence, all data would lie along one curve.

CONSISTENCY

These graphs of iso-electronic sequences are shown to the experts not only to obtain recommendations on ions where there are no data, but also to check the data for consistency and to check for errors. After seeing curves such as those in Figure 6 the experts may ask that some data be checked again or they may wish to revise their recommendations.

Another example of curves brought together to check consistency is illustrated in Figure 7, which shows the rate coefficients for all of the ions of iron.

We found that when the first recommendations on data were made that the curves, such as those in Figure 8 for the ions of cobalt, sometimes crossed - which our experts thought was possible, but unlikely. On looking back on the original recommendations we found that there were often cases where we were using classical scaling to decide on data for a particular ion and there was uncertainty about whether to scale from data lower down in the iso-electronic sequence or higher up in the sequence. When these unlikely crossings occurred we went back and scaled from a different ion, then we checked our curves and if we found that the curves no longer crossed, the second scaling was then preferred to the first.

The example illustrated in Figure 8 can be understood by reference to the iso-electronic sequence for Ca - like ions in Figure 9. The only data available for ions are data on Ti^{2+} and Fe^{6+} which are almost a factor of 2 different from one another. In this circumstance it is natural for the ion Co^{7+} to scale using the data for the ion next in the sequence, Fe^{6+}. However, when scaled from Ti^{2+}, though further away in the sequence we get the curves in Figure 10 and the rates no longer cross. Later we realised that in this particular case the Fe^{6+} data include a large autoionization contribution which do not scale to Co^{7+}.

Before leaving the graphical representation of data, I would like to point out in Figure 11 one case where the empirical law of Lotz, used very widely for electron ionization, is wrong by a large margin. There are many other similar examples; so the Lotz formula can no longer be recommended.

USER INTERFACE

Our final report [15] has presented the data in the form of graphs as in Figures 5 to 11. But in addition, to suit the needs of users we consulted those users in the Culham Laboratory about the format of the data they would most prefer. As a result the data were presented in two additional forms: as parameters to curve fits as in Table 1 and as tables of numbers as in Table 2. So a user can either look at a graph, at numbers in a table or compute the data from a formula on his computer.

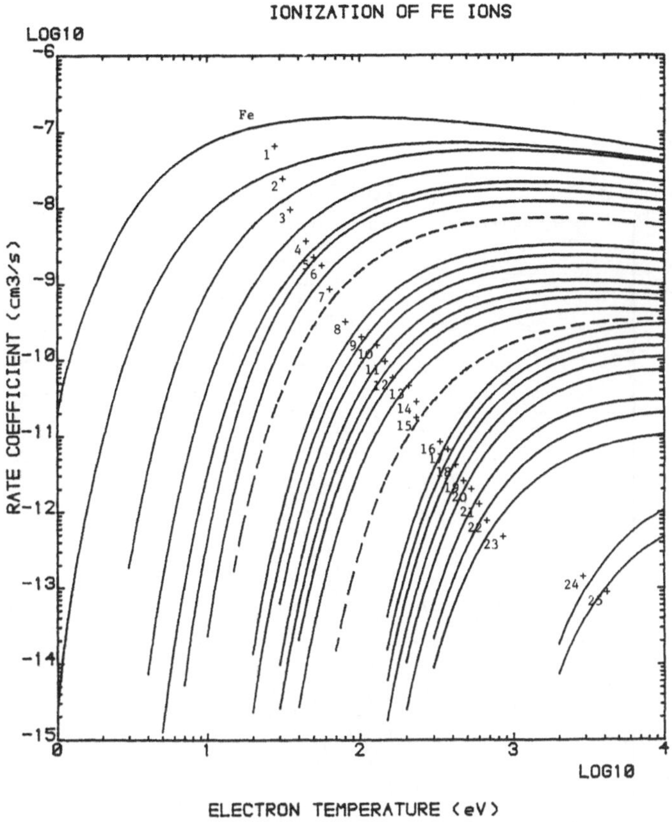

Figure 7

Electron ionization rates for all the ions of iron. The dashed lines are determined using classical scaling. This shows that all of our recommendations are reasonably consistent with one another.

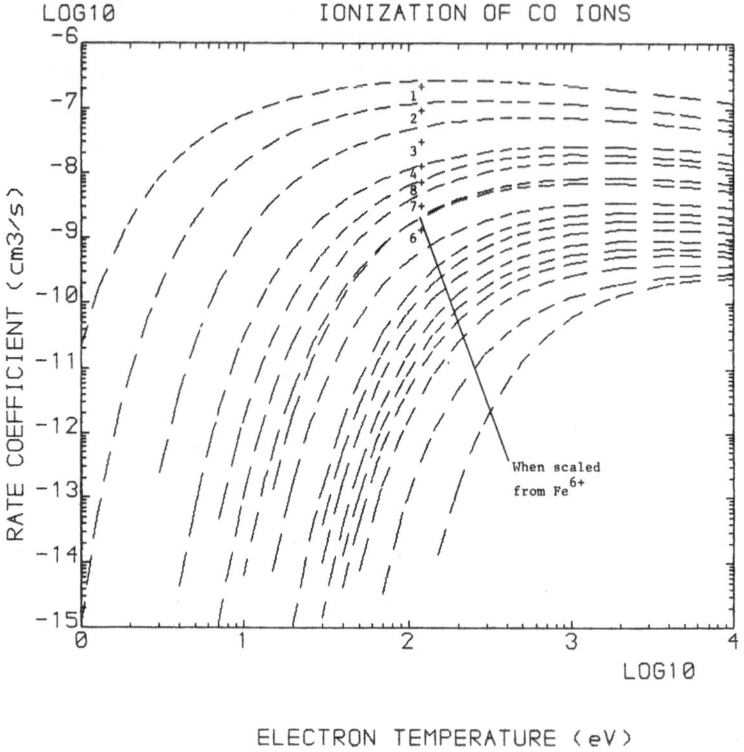

Figure 8

Electron ionization rates first estimated for all ions of Cobalt, showing the unlikely crossing of the rates Co^{6+} and Co^{7+}.

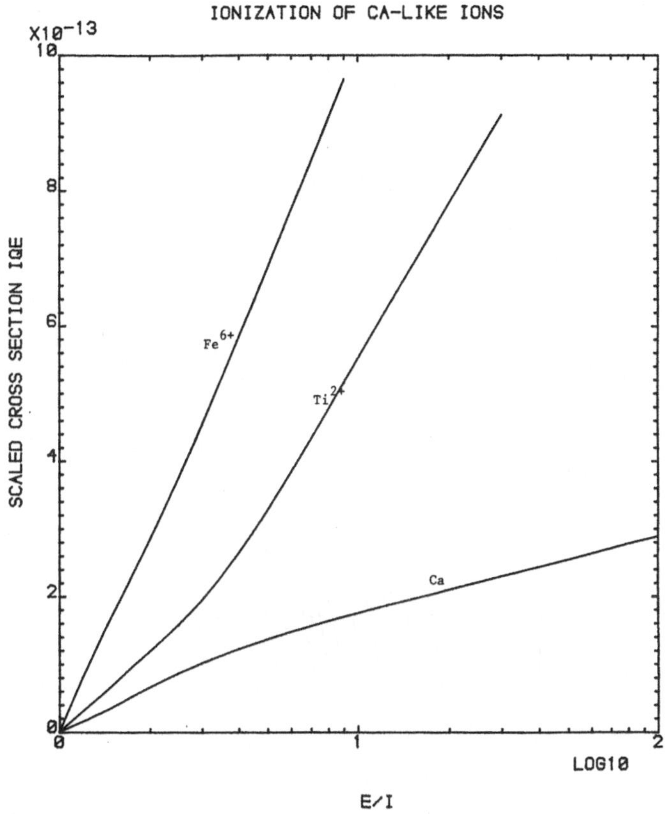

Figure 9

Only data available on the electron ionization of Ca-like ions. The Co^{7+} data in Figure 8 are erroneously scaled from the Fe^{6+} data which includes a large autoionization contribution.

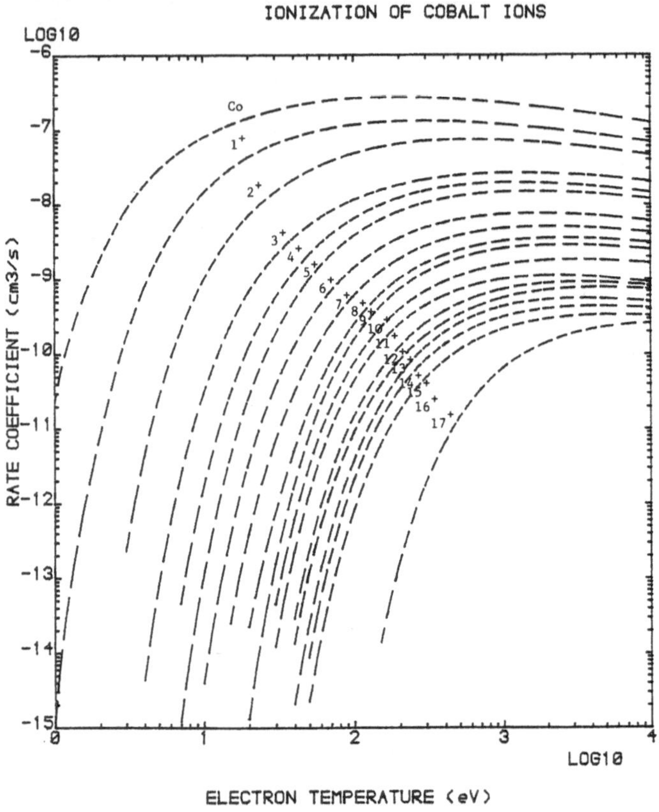

Figure 10

Corrected rates for Co ions with the data for Co^{7+} now scaled from the Ti^{2+} data in Figure 9.

PROCESS : Electron impact ionization

REACTION: Fe(+) + e -> Fe(2+) + 2e

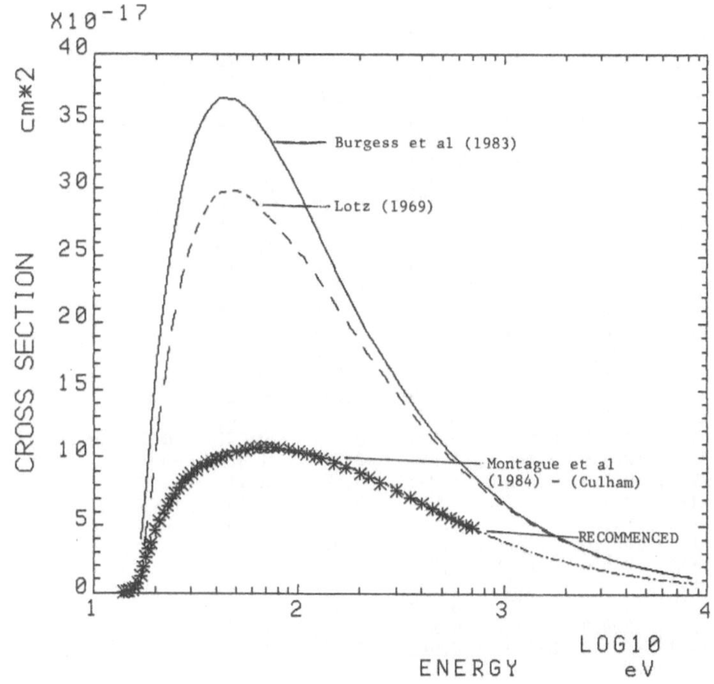

Figure 11

Comparison of recommended data for Fe^+ with empirical data from Lotz and Burgess et al. showing that these empirical data are unreliable in many cases.

Table 1. The parameters I, A and B of formula (1) of the text for the recommended cross sections. The ionisation potential I is given in eV and the parameters A and B are in units of 10-13 eV2 cm2.

SPECIES	I(ev)	A	B_1	B_2	B_3	B_4	B_5	
Fe I	7.9	1.142	-0.920	1.782	-1.694			
Fe II	16.2	2.124	-0.530	-9.617	29.170	-40.810	19.290	
Fe III	30.6	4.420	-2.968	-3.712	1.219	0.044		
Fe IV	54.8	2.258	-0.843	0.622	0.010			
	80.0	6.113	-1.148	1.583	-0.034			E/I > 2.10
Fe V	75.0	2.659	-1.068	-0.396	1.036			
	100.0	6.503	-0.278	-5.663	3.860			E/I > 1.79
Fe VI	99.0	2.888	0.753	-21.240	66.290	-75.330	29.080	
Fe VII	125.0	2.876	-2.321	5.910	-10.306	-1.333	7.738	
Fe VIII	151.1	2.370	0.331	-1.323	-7.778	14.921	-6.092	
Fe IX	233.6	2.856	0.209	0.488	-0.379			
Fe X	262.1	2.457	0.289	0.484	-0.413			
*Fe X	214.0	2.432	-0.676	0.598				
	262.1	1.950	-0.946	11.125	-7.958			E/I > 2.10
Fe XI	290.3	2.034	0.410	0.253	-0.321			
Fe XII	330.8	1.447	0.305	1.596	-1.244			
Fe XIII	361.0	1.336	0.262	0.595	-0.518			
Fe XIV	392.2	1.303	0.149	0.078	-0.123			
Fe XV	457.0	1.133	0.191	-0.150	0.032			
Fe XVI	489.5	0.907	0.105	0.147	-0.073	-0.001		
Fe XVII	1266.0	3.441	1.066	-2.774	1.489			
Fe XVIII	1366.0	2.492	1.043	-0.528	-0.113			
Fe XIX	1463.0	1.912	1.079	-0.035	-0.447			
Fe XX	1583.0	1.633	0.645	-0.101	-0.239			
Fe XXI	1678.0	1.091	0.276	1.052	-0.892			
Fe XXII	1789.0	0.189	0.327	1.679	-1.160			
Fe XXIII	1950.0	0.414	0.097	0.137	-0.144			
Fe XXIV	2046.0	0.289	0.110	-0.441	0.269			
Fe XXV	8829.0	0.525	-0.188	0.489	-0.292			
Fe XXVI	9277.0	0.260	-0.022	0.051	-0.023			

$$\sigma(E) = \frac{1}{IE} \{A \ln (E/I) + \sum_{i=1}^{N} B_i \left(1 - \frac{I}{E}\right)^i\}$$

where E is the incident electron energy
I is the ionization potential
and B_i are the coefficients determined by a least squares fitting procedure

Table 2. Recommended cross sections and rates.

RECOMMENDED CROSS SECTIONS AND RATES

E/I	Fe X -> Fe XI E (eV)	CROSS SECTION (cm2)	* Fe X -> Fe XI E (eV)	CROSS SECTION (cm2)	Fe XI -> Fe XII E (eV)	CROSS SECTION (cm2)
1.25	3.28E+02	7.24E-19	2.68E+02	7.54E-19	3.63E+02	5.16E-19
1.50	3.93E+02	1.10E-18	3.21E+02	1.20E-18	4.35E+02	7.73E-19
1.75	4.59E+02	1.29E-18	3.75E+02	1.47E-18	5.08E+02	9.05E-19
2.00	5.24E+02	1.40E-18	4.28E+02	1.63E-18	5.81E+02	9.72E-19
2.25	5.90E+02	1.44E-18	4.82E+02	1.83E-18	6.53E+02	1.00E-18
2.50	6.55E+02	1.46E-18	5.35E+02	1.96E-18	7.26E+02	1.01E-18
2.75	7.21E+02	1.46E-18	5.89E+02	2.02E-18	7.98E+02	1.01E-18
3.00	7.86E+02	1.45E-18	6.42E+02	2.04E-18	8.71E+02	9.99E-19
3.50	9.17E+02	1.41E-18	7.49E+02	2.01E-18	1.02E+03	9.67E-19
4.00	1.05E+03	1.35E-18	8.56E+02	1.94E-18	1.16E+03	9.30E-19
4.50	1.18E+03	1.30E-18	9.63E+02	1.85E-18	1.31E+03	8.91E-19
5.00	1.31E+03	1.25E-18	1.07E+03	1.76E-18	1.45E+03	8.54E-19
6.00	1.57E+03	1.15E-18	1.28E+03	1.60E-18	1.74E+03	7.86E-19
7.00	1.83E+03	1.07E-18	1.50E+03	1.46E-18	2.03E+03	7.28E-19
8.00	2.10E+03	9.93E-19	1.71E+03	1.34E-18	2.32E+03	6.77E-19
9.00	2.36E+03	9.30E-19	1.93E+03	1.24E-18	2.61E+03	6.34E-19
10.00	2.62E+03	8.75E-19	2.14E+03	1.15E-18	2.90E+03	5.96E-19
12.50	3.28E+03	7.64E-19	2.68E+03	9.82E-19	3.63E+03	5.20E-19
15.00	3.93E+03	6.80E-19	3.21E+03	8.60E-19	4.35E+03	4.63E-19
20.00	5.24E+03	5.62E-19	4.28E+03	6.93E-19	5.81E+03	3.82E-19
30.00	7.86E+03	4.23E-19	6.42E+03	5.08E-19	8.71E+03	2.87E-19
40.00	1.05E+04	3.43E-19	8.56E+03	4.05E-19	1.16E+04	2.33E-19
50.00	1.31E+04	2.90E-19	1.07E+04	3.39E-19	1.45E+04	1.97E-19
75.00	1.97E+04	2.13E-19	1.61E+04	2.44E-19	2.18E+04	1.44E-19
100.00	2.62E+04	1.70E-19	2.14E+04	1.93E-19	2.90E+04	1.15E-19

T (eV)	T (Deg.K)	RATE COEFF. (cm3/s)	T (Deg.K)	RATE COEFF. (cm3/s)	T (Deg.K)	RATE COEFF. (cm3/s)
1.00	1.16E+04	-	1.16E+04	-	1.16E+04	-
2.00	2.32E+04	4.58E-67	2.32E+04	1.24E-56	2.32E+04	-
3.00	3.48E+04	5.24E-48	3.48E+04	4.66E-41	3.48E+04	3.14E-52
4.00	4.64E+04	1.84E-38	4.64E+04	2.99E-33	4.64E+04	1.16E-41
5.00	5.80E+04	1.01E-32	5.80E+04	1.48E-28	5.80E+04	2.60E-35
7.00	8.12E+04	3.80E-26	8.12E+04	3.58E-23	8.12E+04	4.89E-28
10.00	1.16E+05	3.40E-21	1.16E+05	4.10E-19	1.16E+05	1.47E-22
15.00	1.74E+05	2.56E-17	1.74E+05	6.26E-16	1.74E+05	2.82E-18
20.00	2.32E+05	2.30E-15	2.32E+05	2.54E-14	2.32E+05	4.04E-16
30.00	3.48E+05	2.16E-13	3.48E+05	1.08E-12	3.48E+05	6.07E-14
40.00	4.64E+05	2.15E-12	4.64E+05	7.32E-12	4.64E+05	7.66E-13
50.00	5.80E+05	8.69E-12	5.80E+05	2.35E-11	5.80E+05	3.56E-12
70.00	8.12E+05	4.38E-11	8.12E+05	9.19E-11	8.12E+05	2.11E-11
100.00	1.16E+06	1.51E-10	1.16E+06	2.66E-10	1.16E+06	8.17E-11
150.00	1.74E+06	4.02E-10	1.74E+06	6.31E-10	1.74E+06	2.39E-10
200.00	2.32E+06	6.62E-10	2.32E+06	9.87E-10	2.32E+06	4.14E-10
300.00	3.48E+06	1.10E-09	3.48E+06	1.55E-09	3.48E+06	7.19E-10
400.00	4.64E+06	1.41E-09	4.64E+06	1.94E-09	4.64E+06	9.49E-10
500.00	5.80E+06	1.64E-09	5.80E+06	2.21E-09	5.80E+06	1.12E-09
700.00	8.12E+06	1.93E-09	8.12E+06	2.52E-09	8.12E+06	1.34E-09
1000.00	1.16E+07	2.17E-09	1.16E+07	2.72E-09	1.16E+07	1.52E-09
2000.00	2.32E+07	2.37E-09	2.32E+07	2.79E-09	2.32E+07	1.70E-09
4000.00	4.64E+07	2.32E-09	4.64E+07	2.57E-09	4.64E+07	1.68E-09
5000.00	5.80E+07	2.27E-09	5.80E+07	2.47E-09	5.80E+07	1.65E-09
10000.00	1.16E+08	2.03E-09	1.16E+08	2.10E-09	1.16E+08	1.48E-09

ACKNOWLEDGEMENT

I wish to acknowledge support in part from the United Kingdom Atomic Energy Authority, Culham Laboratory. The Author is also heavily indebted to Dr Marguerite Lennon who did most of the work leading to the production of this paper, including all the graphs.

REFERENCES

1. Bates, D.R. and Unwin, J.J., June 1938, Recombination in the Upper Atmosphere, MSc Dissertation, The Queen's University of Belfast.
2. Thompson, J., 1876, On an Integrating Machine having a new Kinematic Principle, Proc.Roy.Soc. XXIV, 262-8.
3. Thompson, W., 1878, Harmonic Analyser, Proc.Roy.Soc. XXVII, 371-3.
4. Hartree, D.R. and Porter, A., 1935, The construction and operation of a model differential analyser, Mem.Man.Lit.Phil.Soc. 79, 51.
5. Massey, H.S.W., Wylie, J., Buckingham, R.A. and Sullivan, R., 1938, A Small Scale Differential Analyser - its Construction and Operation, Proc.Roy.Irish Acad. XLV, 1-21.
6. Bates, D.R., Ledsham, K. and Stewart, A.L., 1953, Wave functions of the hydrogen molecular ion, Phil.Trans.A246, 215-240.
7. Burke, P.G., Burke, V.M., Percival, I.C. and McCarroll, R., 1962, Electron scattering by atomic hydrogen in the 1s, 2s or 2p state, Proc.Phys.Soc. 80, 413-421.
8. Boyle, J., McDonough, W.R., O'Hara, H. and Smith, F.J., 1977, An On-line Numerical Data System, Program 11, 35-40.
9. Online Information, 7th International Meeting, London, December 1983, Learned Information Limited, Oxford.
10. Pease, R.S., 1978, Towards a Controlled Nuclear Fusion Reactor, IAEA Bulletin 20, No.6, 9-12.
11. Atomic and Molecular Data for Fusion. Proceedings of an advisory group meeting on Atomic and Molecular Data for Fusion organised by the International Atomic Energy Agency held at the UKAEA Culham Laboratory, Abingdon, UK, 1-5 November 1976, IAEA - 199, 587 pages.
12. International Bulletin on Atomic and Molecular Data for Fusion, published quarterly by the International Atomic Energy Agency, Vienna.
13. CIAMDA 80. An Index to the Literature on Atomic and Molecular Collisions Relevant to Fusion Research. International Atomic Energy Agency, Vienna 1980.
14. Bell, K.L., Gilbody, H.B., Hughes, J.G., and Kingston, A.E., 1982, Atomic and Molecular Data for Fusion, Part 1. Recommended cross sections and rates for electron ionisation of light atoms and ions. UKAEA Report CLM-R216.

INDEX